THE

EARTH AND MAN:

LECTURES ON

COMPARATIVE PHYSICAL GEOGRAPHY,

IN ITS

RELATION TO THE HISTORY OF MANKIND.

BY

ARNOLD GUYOT,

LATE PROF. OF PHYSICAL GEOGRAPHY AND HISTORY, AT NEUCHATEL, SWITZERLAND.

TRANSLATED FROM THE FRENCH, BY

C. C. FELTON,

PROFESSOR IN HARVARD UNIVERSITY.

Our Earth is a star among the stars; and should not we, who are on it, prepare ourselves by it for the contemplation of the Universe and its Author? — CARL RITTER.

FOURTEENTH THOUSAND.

BOSTON:
GOULD AND LINCOLN.
NEW YORK: SHELDON AND COMPANY.
CINCINNATI: GEO. S. BLANCHARD.
1860.

PRINTED BY
GEORGE C. RAND & AVERY.

Stereotyped by
HOBART & ROBBINS;
NEW ENGLAND TYPE AND STEREOTYPE FOUNDERY,
BOSTON.

To C. C. FELTON, Esq.,

Professor in Harvard University, Cambridge.

My Dear Sir,

It is to your friendship that I owe the idea and the possibility of publishing this little work in a language not my own. With rare kindness, and a disinterestedness still more rare, you have placed at my disposal your hours of leisure and your skilful and indefatigable pen. The book is yours already, and I but renew your title to it by begging you to accept this DEDICATION, as a testimony of my heart-felt gratitude, and at the same time as a souvenir of the long but pleasant hours of labor which you have so kindly shared with me to the last.

THE AUTHOR

PREFACE.

The lectures contained in the volume here offered to the public were delivered, by invitation, in French, between the 17th of January and the 24th of February, of the present year. One of the halls of the Lowell Institute, in Boston, was placed, for that purpose, at the author's disposal, by the liberality of the Trustee, John A. Lowell, Esq. They were spoken with the help only of a few notes, and were not intended, at the time, for the press. But the publication having been desired by some friends, and requested by the editors of the *Boston Daily Traveller*, for the columns of that excellent journal, the author determined to write out, the next morning, the lecture of the evening before. These rapid pages, translated, from day to day, by Mr. C. C Felton, Professor in Harvard University, are collected and reprinted in the present volume. Neither time nor circumstances have permitted any important alterations; the only material additions are found in the first lecture, the last part of which did not appear in the journal, and, at the beginning of the eighth, the portion which treats of the marine currents. This subject, although announced in the programme of the course, it was found necessary, for want of time, to pass over in silence. As to the rest, the lectures have retained their original cast, notwithstanding the incongruity which sometimes happens, of bringing several different subjects into the same discourse.

This brief history of the present book will place the reader in a position to form a just opinion of the work, and perhaps will induce him to extend to it some indulgence.

It will, moreover, be readily understood, that oral instruction is naturally clothed in forms appropriate to itself, which are not those of a systematic and didactic exposition, such as is required by a book intended only for reading, or for the silent study of the closet. In the opinion of the author, it should bring out in strong relief, even by venturing a dash of the pencil somewhat bold, the essential traits of the subject, in order to fix and deepen the impression, while the secondary features are thrown into the shade. Truth, far from losing by this mode, will gain the advantage of being grasped in a manner at once more distinct and more correct. For nothing is less indispensable to true science, — may the reader of these pages find it so, — than the scholastic and doctoral robe, which is too often unnecessarily worn.

This little work is not, then, a treatise on the subject indicated by its title. The author would wish to consider this unforeseen publication only as the forerunner of a more complete work, the materials of which, gradually collected during long years of study, and still daily accumulating, he hopes to arrange, and work out more at leisure, if not in the same form, at least in the same spirit. However, he is confident that the man of science will find, in this first sketch, the traces of serious and matured studies.

Numerous quotations and references were incompatible with the form of these discourses. The facts, properly so called, are drawn from the common domain of science; and as to the results that have been deduced from their combination, the author willingly leaves to men versed in the subject the task of distinguishing those which may be regarded as constituting a progress in knowledge of the creation, and of its relations to man.

There are, however, three names so closely connected with the history of the science to which this volume is devoted, and with the past studies of the author, that he feels bound to mention them here Humboldt, Ritter, and Steffens, are the three great minds who have breathed a new life into the science of the physical and moral world. The scientific life of the author opened under the full radiance of the light they spread around them, and it is with a sentiment of filial piety that he delights to recall this connection, and to render to them his public homage.

Notwithstanding the praiseworthy care the publishers of this volume have taken to provide it with the maps and drawings necessary to understand the text, the reader will perhaps desire more. He will find them in the Physical Atlas of Berghaus, the most excellent, and almost the only work of the kind, or in the English publications based on it, by Johnston of Edinburgh, by A. Petermann of London, and others. The explanatory pages give the information necessary for the plates that accompany these sheets. For their execution on stone, the author deems himself happy in having been able to avail himself of the talents of an artist so able and obliging as M. Sonrel.

Besides Prof. Felton, who has read all the proof-sheets, the author returns his sincere acknowledgments to Professors Agassiz, Peirce, and Gray, who have had the goodness to revise portions of them.

Few subjects seem more worthy to occupy thoughtful minds, than the contemplation of the grand harmonies of nature and history. The spectacle of the good and the beautiful in nature, reflecting everywhere the idea of the Creator, calms and refreshes the soul. The view of the hand of Providence, guiding the chariot of human destinies, reässures and strengthens our faith. May these unpretending sheets, launched upon the sea of publicity, reach those who feel the need of both, and by them be kindly received.

CAMBRIDGE MASS., *May* 1, 1849

PREFACE TO THE SECOND EDITION.

THE marked favor with which the public, here and abroad,* have received this essay, imposed on the author the duty of carefully revising it. But the time of its first appearance is so recent, that no important alterations are to be made. Except a few additions of facts quite recently acquired for science, particularly in the paragraph relating to the sub-marine relief of the basin of the oceans, the work has remained, in substance, what it was. Not so with the translation. The translator, exercising a severer criticism than the reader upon his own work, has carefully revised, improved, and corrected it; and the author seizes this occasion to repeat his thanks for this fresh endeavor to render these sheets more worthy of the public approbation.

If there is any reward worthy of desire to him who communicates his thoughts to his fellow-men, it is that of meeting, in the midst of throngs preöccupied with so many diversified cares, an echo and sympathy. This gratification has not been wanting to the author, and he recalls with gratitude the numerous testimonies he has received from so many quarters. He finds in these manifestations an encouragement to continue his work, and to prepare a second volume, on the Historical Development of Humanity, which he considers as the necessary complement to the present.

CAMBRIDGE, *July*, 1850.

* This volume has been republished in London, by Bentley, and an edition in French is about to appear at Paris. A mutilated edition, called "revised," has also been published by E. Gover, Sen., London, in which many passages, amounting to over thirty pages of the original edition, and essential to the continuity of the argument, or containing conclusions, have been suppressed; additions have been inserted expressing views not advanced by the author; alterations have been made, in exceedingly bad English, all without the least intimation in the preface. Against the two first, the author protests; against the last, the translator.

EXPLANATION OF THE PLATES AND FIGURES

PLATE I. Physical Map of the World.

This map, in Mercator's projection, is intended to enable the eye to seize at a glance the great physical features of the surface of the globe. To this end, each particular is indicated by a different color. The color of the ocean forms a ground which clearly defines and brings out the characteristic forms of the continents. The bands of white lines which cross them indicate the course of the marine currents, according to the Physical Map of Berghaus. The arrows mark their direction. In the continents, three colors distinguish the three principal forms of relief from each other. The green tint marks the low lands; the white, the more elevated parts and the table lands; the brown, the systems of mountains, the borders of the table lands, and the slopes in general. It is easy thus to form by a single glance an idea of the general features of relief of the different countries of the earth. The dotted lines which cross the map are, beginning at the top, the Arctic circle, the tropic of Cancer, and the tropic of Capricorn. The entirely straight line is the Equator. The latitudes are marked in the margin by a line for every 15°. The longitudes in the same way, by 15° East and West from the meridian of Paris. The two winding lines in the northern half of the map are the isothermal lines of zero Centigrade or 32° Fahrenheit, and of 15° Centigrade or 59° Fahrenheit. All the places situated on these lines, having the same mean annual temperature, set in a clear light the difference of climate between the opposite coasts of the continents, while referring it to the true causes. Every letter has been omitted from this little map, which is intended to be a physical picture, and to speak to the eye. The scale, moreover, scarcely allowed their insertion, and the great features which it represents are so well known that there is no need of naming them.

PLATES II. AND III. (pp. 60, 61.)

These plates contain a series of ideal profiles intended to illustrate the general laws of relief of the continents. The profiles comprise each a transverse zone rather than a simple line, which often would have answered but imperfectly the proposed end. The relation of the heights to the horizontal distances would have to be magnified about one hundred times. The numbers placed in the margin, indicate the heights in thousands of feet. The letters placed at the top of the vertical dotted lines, are the initials of the names of the principal points contained in the tables; when the same initial is repeated, the second in the order of the table is marked thus: ('). The profiles of the massive and entire parts of the continents, comprising the plains and table lands, are distinguished from the heights which surround them, or the mountain chains, by a particular line, by different hatchings, and deeper shading. The peaks, which are merely indicated above the base line, without being connected, are either mountains situated outside of the zone, followed by the profile, as the Carpathian and Mont-Blanc, in Europe, Plate II. profile 5; or volcanic peaks, isolated, not affecting the general relief, as the Erdschich, in Asia Minor, Plate II. profile 4; the Ararat, in Armenia, Plate III. profile 1. Plate II. profile 3; the St. Elias, in North America.

Plate II. comprises 7 profiles across the three principal continents of the Old World, in the direction from north to south. In profile 1, Eastern Asia, and profile 6, Africa, the line of horizontal distances being too considerable to be taken into the frame, the profiles are interrupted to indicate that one portion of the horizontal line has been suppressed. In the profile of Eastern Asia the portion omitted is almost equal to the less section. In Africa it is much larger still.

Plate III. comprises the profiles of the New World, from east to west. No. 2, passing along the line of the Antilles, is necessarily broken. But the gradual increase of the reliefs and their disposition prove that this line ought to be considered in reality continuous, although at some points it is covered by the waters of the ocean. The profiles are arranged in the plates in such a manner as to show, at once, in the vertical line, the increase of the reliefs from west to east in the Old World, and from north to south in the New World. The text itself makes further explanation unnecessary.

PLATE IV. Map of the Distribution of Rain.

This map, taken from the Physical Atlas of Berghaus, shows the distribution of rain on the surface of the globe. The deeper the color, the greater is the quantity of rain-water indicated; the deserts are left in white. North and south of the tropics, which are marked by dotted lines, are the regions of continuous, but not abundant rains. Between the tropics, the region of periodical and copious rains. A little north of the Equator a deeper shaded strip indicates the region of calms, where daily thunder storms cause almost throughout the year the fall of a considerable quantity of water.

PLATES V. AND VI.,

Intended to illustrate the law of the degeneration of the human type in leaving the central region of Western Asia, comprise 16 portraits, all drawn from nature, and taken from the plates of the "Animal Kingdom" of Cuvier, wherever a different source is not indicated.

Plate V. *First Series.* From the central regions of Western Asia, to the extremity of Africa, through Arabia and the eastern coast.

No. 1. A Circassian, belonging to the suite of the Persian Ambassador; drawn from the life, at Paris, in 1823, by M. A. Colin.

No. 2. An Arab of Algiers, of the Mozabite tribe; drawn from life, under the direction of Mr. Milne Edwards, by A. Lordon.

No. 3. A negro of Mozambique, on the south-east coast of Africa; drawn from life, in Brazil, by Rugendas.

No. 4. Joshua Makoniane, an old Bassouto warrior, a convert to Christianity, drawn from life by Mr. Maeder, of the French Mission to South Africa. Journal des Missions Evangeliques de Paris, Vol. XX.

Second Series. From Europe to tropical Africa, by the western coast.

No. 5. Portrait of Captain Cook, painted by Dauce, in the gallery of the Naval Hospital at Greenwich. Geographical Almanac of Berghaus.

No. 6. A Cabyle of Flissa, in Algeria; drawn from life by A. Lordon.

No. 7. Senegal Chief, after an unpublished drawing by an officer in the expedition of Captain Laplace.

No. 8. A Negro of Congo; drawn from nature by Rugendas, Voyage Pittoresque au Brésil.

Plate VI. *First Series.*

No. 1. Mongolian type, portrait of one of the Siamese twins, seen in Europe in 1830, after a drawing made from nature, at Paris.

No. 2. Malay, belonging to the group of the Koutousoff Smolensky, from a plate in the work of Choris, Voyage du Rurick.

No. 3. New Holland. Portrait of Onrou-Mare, a warrior of the tribe of the Gwea-Gul, from the Atlas du Voyage aux Terres Australes.

No. 4. A woman of Van Diemen, from l'Atlas de l'Astrolabe.

Second Series. America, from the sources of the Missouri to Terra del Fuego; and the Polar variety.

No. 5. Oto Indian, portrait taken from the Travels of Prince Maximilian of Neuwied.

No. 6. Coroado Indian, from the banks of the Rio Xipoto, one of the tributaries of the Rio Pomba, in tropical South America, after a portrait published by Spix and Martius.

No. 7. An inhabitant of Terra del Fuego; Univers Pittoresque.

No. 8. Inhabitant of the Aleutian Islands, after Choris; Voyage of Kotzebue.

Fig. 1. page 43. Land Hemisphere, and Water Hemisphere.

Fig. 2. page 106. Europe at the Silurian Epoch.

Fig. 3. page 108. America at the Coal Epoch.

Fig. 4. page 111. Europe at the Tertiary Epoch.

These three last maps, intended to show the gradual increase of the dry lands, do not so much indicate the real contours of the lands existing at those epochs, — this would be impossible, — as the portions which have not since been covered by the waters of the ocean. The white portions are the only dry land. All the portions in ruled lines are under water; but the existing contours of the continents are represented by means of a lighter shade, as a point of comparison. The maps Fig. 2, and Fig. 4, have been constructed after the geological maps of Elie de Beaumont, (in Beant géologie,) Boué, and Dechen; the map Fig. 3, after the geological map of the United States, by Mr. James Hall, completed by that of Sir Charles Lyell and the geological map of the world, by Boué.

CONTENTS.

LECTURE I.

LECTURE II.

LECTURE III.

LECTURE IV.

LECTURE V.

LECTURE VI.

LECTURE VII.

LECTURE VIII.

LECTURE IX.

LECTURE X.

LECTURE XI.

LECTURE XII.

THE

EARTH AND MAN.

LECTURE I.

Subject of the course — What should be understood by Geography — Definition of Physical Geography — The life of the globe — Importance of the geographical forms of contour and relief, and of their relative situation — The Earth as the theatre of human societies — Different parts performed by the continents in history — Asia, Europe, America — Inquiry into the analogies of the general forms of the continents.

LADIES AND GENTLEMEN : —

In asking your attention to a few scientific discourses, in a language not your own, I have not disguised from myself that this circumstance is perhaps a source of embarrassment for some of you, as it certainly is for me. In the communion of mind with mind, in the mutual interchange of ideas, the first condition necessary for establishing between him who speaks and those that hear, the sympathetic harmony which makes its charm, is, that the word shall reach the understanding without obstacle and without effort.

In my favor you have made the sacrifice of your language. I need not tell you, that, on my part, I will do all in my power to render that sacrifice less irksome, and I shall always be desirous of giving to those who will do me the favor to ask it, all the explanations which they can require.

The subject to which I propose to call your attention, is Comparative Physical Geography, considered in its relations to the history and the destinies of mankind. But the term geography has been applied to such different things, the use, the misuse rather to which it has been subjected, has rendered it so elastic and ill-defined, that, in order to prevent misconception, I must first of all explain to you what I understand by *Geography*.

If, preserving the etymological sense of the word geography, we should, with many authors, undertake to limit this study to a simple description of the surface of the globe and of the beings which are found there, we must at once renounce the idea of calling it by the name of science, in the lofty sense of this word. To describe, without rising to the causes, or descending to the consequences, is no more science, than merely and simply to relate a fact of which one has been a witness. The geographer, who thus understands his study, seems to make as little of geography as the chronicler of history. It would be easy to show that even the power of describing well ought to be denied him; for if he renounces the study of the laws which have presided over the creation, over the disposition of the terrestrial individuals in their different orders: if he will take no account of those which have given birth to the phenomena that he wishes to describe, soon, overwhelmed beneath the mass of details, of whose relative value he is ignorant, without a guide and without a rule to make a judicious choice in the midst of this infinite variety of partial observations, he remains incapable of mastering them, of grouping them in such a manner as to bring prominently forward those which must give character to the whole, and thus dooms himself to a barren confusion at least; happy, if, in place of a faithful picture of nature, he does not finally profess to give us, as such, the strangest caricature.

No! Geography — and I regret here that usage forbids me to employ the most suitable word, *Geology*, to designate the general science of which I speak — Geography ought to be something different from a mere description. It should not only describe, it should compare, it should interpret, it should rise to the *how* and the *wherefore* of the phenomena which it describes. It is not enough for it coldly to *anatomize* the globe, by merely taking cognizance of the arrangement of the various parts which constitute it. It must endeavor to seize those incessant mutual actions of the different portions of physical nature upon each other, of inorganic nature upon organized beings, upon man in particular, and upon the successive development of human societies, in a word, studying the reciprocal action of all these forces, the perpetual play of which constitutes what might be called the life of the globe, it should, if I may venture to say so, inquire into its physiology. To understand it in any other way, is to deprive geography of its vital principle; is to make it a collection of partial, unmeaning facts, is to fasten upon it forever that character of dryness, for which it has so often and so justly been reproached. For what is dryness in a science, except the absence of those principles, of those ideas, of those general results, by which well-constituted minds are nurtured?

Physical geography, therefore, ought to be, not only the description of our earth, but the physical science of the globe, or the science of the general phenomena *of the present life of the globe, in reference to their connection and their mutual dependence.*

This is the geography of Humboldt and of Ritter.

But I speak of *the life of the globe*, of the *physiology* of the great terrestrial forms! These terms may perhaps seem here to be misapplied.

I ask your permission to justify them, for I cannot find better, to express what appears to me to be the truth.

Far from me the idea of attempting to assimilate this general life of the inorganic nature of the globe to the individual life of the plant or the animal, as some unwise philosophers have done. I know well the wide distance which separates inorganic from organized nature. I will even go further than is ordinarily done, and I will say that there is an impassable chasm between the mineral and the plant, between the plant and the animal, an impassable chasm between the animal and the man. But this nature, represented as *dead*, and contrasted in common language with *living* nature, because it has not the same life with the animal or the plant, is it then bereft of all life? If it has not life, we must acknowledge that it has at least the appearances of life. Has it not motion in the water which streams and gushes over the surface of the continents, or which tosses in the bosom of the seas?—in the winds which course with terrible rapidity and sweep the soil that we tread under our feet, covering it with ruins? Has it not its sympathies and antipathies in those mysterious elective affinities of the different molecules of matter which chemistry investigates? Has it not the powerful attractions of bodies to each other, which govern the motions of the stars scattered in the immensity of space, and keep them in an admirable harmony? Do we not see, and always with a secret astonishment, the magnetic needle agitated at the approach of a particle of iron, and leaping under the fire of the Northern light? Place any material body whatsoever by the side of another, do they not immediately enter into relations of interchange, of molecular attraction, of electricity, of magnetism? The disturbance of the equilibrium at one point induces another elsewhere, and the movement is propagated to infinity. And what will it be, if we rise to the contemplation of all the phenomena of this order together, exhibited by a vast country, by an entire continent?

Thus, in inorganic nature likewise, all is acting, all is changing, all is undergoing transformation. Doubtless this is not the life of the organized being, the life of the animal; but is not this assemblage of phenomena also a life? If, taking life in its most simple aspect, we define it as a *mutual exchange of relations*, we cannot refuse this name to those lively actions and reactions, to that perpetual play of the forces of matter, of which we are every day the witnesses. Yes, gentlemen, it is indeed life, but undoubtedly in a very inferior order of things.

It is life; the thousand voices of nature which make themselves heard around us, and which in so many ways betray that incessant and prodigious activity, proclaim it so loudly that we cannot shut our ears to their language.

This general life, this physical and chemical life, belongs to all matter. It is the basis of the existence of all superior beings, not as the source, but as the condition. It is in the plant, it is in the animal; only here it is subservient to a principle of higher life of a spiritual nature, of a principle of unity, the mysterious force of which, referring all to a centre, modifies it, controls it, and organizes it, for the benefit of an individual.

Now it is precisely this *internal* principle of unity belonging to organized nature, which is wanting in individuals of inorganic nature; and that is the difference.

In inorganic nature, the bodies are only simple aggregations of parts, homogeneous or heterogeneous, and differing among themselves, the combination of which seems to be accidental. Nevertheless, to say nothing of the law that assigns to each species of mineral a particular form of crystallization, we see that every aggregation, fortuitous in appearance, may constitute a whole, with limits, and a determinate form, which, without having anything of absolute necessity, gives to it, however, the first lineaments of individuality. Such are the various geographical regions, the islands, the peninsulas, the continents; the Antilles, for example, England, Italy, Asia, Europe, North America. Each of these terrestrial masses, considered as a whole, as an individual, has a particular disposition of its parts, of the forms which belong only to it, a situation relatively to the rays of the sun, and with respect to the seas or the neighboring masses, not found identically repeated in any other.

All these various causes excite and combine, in a manner infinitely varied, the play of the physical forces inherent in the matter composing them, and secure to each a climate, a vegetation, and animal life; in a word, an assemblage of physical characters and functions peculiar to it, and really giving it something of individuality.

It is in this sense that we shall speak of the great geographical individuals, that we shall be able to define them, to indicate their characters, to mark their differences; in a word, to apply to them that comparative study, without which there is no true science. But let us not forget that these individuals have the cause of their existence, not *within*, like organized beings, but *without*, in the very circumstances of their aggregation. Hence, gentlemen, the great importance of external form; the importance of the geographical forms of contour, of relief of the terrestrial surface; of the relations of size, of extent, of relative position. In considering them simply in a geological point of view, it may appear quite accidental that such a plain should or should not have risen from the bosom of the waters; that such a mountain rises at this place or that; that such a continent should be cut up into peninsulas, or piled into a compact mass, accompanied by, or deprived of, islands. When, finally, we reflect that a depression of a few hundred feet, which would make no change in the essential forms of the solid mass of the globe, would cause a great part of Asia and of Europe to disappear beneath the water of the oceans, and would reduce America to a few large islands, we might be led to the conclusion that the external shape of the continents has but an inconsiderable importance.

But, in physics, neither of these circumstances is unimportant. Simple examples, without further demonstration, will be sufficient to set this in a clear light.

Is the question of the forms of contour? Nothing characterizes Europe better than the variety of its indentations, of its peninsulas, of its islands. Suppose, for a moment, that beautiful Italy, Greece with its entire Archipelago, were added to the central mass of the continent, and augmented Germany or Russia by the number of square miles they contain; this change of form would not give us another Germany, but we should have an Italy and a Greece the less. Unite with the body of Europe all its islands and peninsulas into one compact mass, and instead of this continent, so rich in various elements, you will have a New Holland with all its uniformity.

Do we look to the forms of relief, of height? Is it a matter of indifference whether an entire country is lifted into the dry and cold regions of the atmosphere, like the central table land of Asia, or is placed on the level of the ocean? See, under the same sky, the warm and fertile plains of Hindoostan, adorned with the brilliant vegetation of the tropics, and the cold and desert plateaus of Upper Tubet; compare the burning region of Vera Cruz and its fevers, with the lofty plains of Mexico and its perpetual spring; the immense forests of the Amazon, where vegetation puts forth all its splendors, and the desolate paramos of the summits of the Andes, and you have the answer.

And the relative position? Do not the three peninsulas of the south of Europe owe to their position their mild and soft climate, their lovely landscape, their numerous relations, and their common life? Is it not to their situation that the two great peninsulas of India are indebted for their rich nature, and the conspicuous part one of them, at least, has played in all ages? Place them on the north of their continents, Italy and Greece become Scandinavia, and India a Kamtschatka.

All Europe is indebted for its temperate atmosphere to its position relatively to the great marine and atmospheric currents, and to the vicinity of the burning regions of Africa. Place it at the east of Asia, it will be only a frozen peninsula.

Suppose the Andes, transferred to the eastern coast of South America, hindered the trade wind from bearing the vapors of the ocean into the interior of the continent, and the plains of the Amazon and of Paraguay would be nothing but a desert.

In the same manner, if the Rocky Mountains bordered the eastern coast of North America, and closed against the nations of the East and of Europe the entrance to the rich valley of the Mississippi; or if this immense chain extended from east to west across the northern part of this continent, and barred the passage of the polar winds, which now rush unobstructed over these vast plains; — let us say even less: if, preserving all the great present features of this continent, we suppose only that the interior plains were slightly inclined towards the north, and that the Mississippi emptied into the Frozen Ocean, who does not see that, in these various cases, the relations of warmth and moisture, the climate, in a word, and with it the vegetation and the animal world, would undergo the most important modifications, and that these changes of form and of relative position would have an influence greater still upon the destinies of human societies, both in the present and in the future?

It would be easy to multiply examples; but I do not wish to anticipate the results that will be brought out by the more exact study of these phenomena, which we are about to undertake. It is enough for me to have opened a view of the important part performed by all these physical circumstances, and the necessity of studying them with the most scrupulous care.

Let us not, then, despise the study of these outward forms, the influence of which is so evident. They are *everything* in this class of things.

We shall see all the great phenomena of the physical and individual life of the continents, and their functions in the great whole, flowing from the *forms* and the *relative situation* of the great terrestrial masses, placed under the influence of the general forces of nature.

But, gentlemen, it is not enough to have seized, in this point of view, entirely physical as yet, the functions of the great masses of the continents. They have others, still more important, which, if rightly understood, ought to be considered as the final end for which they have received their existence. To understand and appreciate them at their full value, to study them in their true point of view, we must rise to a higher position. We must elevate ourselves to the moral world to understand the physical world; the physical world has no meaning except by and for the moral world.

It is, in fact, the universal law of all that exists in finite nature, not to have, in itself, either the reason or the entire aim of its own existence. Every being exists, not only for itself, but forms necessarily a portion of a great whole, of which the plan and the idea go infinitely beyond it, and in which it is destined to play a part. Thus inorganic nature exists, not only for itself, but to serve as a basis for the life of the plant and the animal; and in their service it performs functions of a kind greatly superior to those assigned to it by the laws which are purely physical and chemical. In the same manner, all nature, our globe, admirable as is its arrangement, is not the final end of creation; but it is the condition of the existence of man. It answers as an instrument by which his education is accomplished, and performs, in his service, functions more exalted and more noble than its own nature, and for which it was made. The superior being then solicits, so to speak, the creation of the inferior being, and associates it to his own functions; and it is correct to say that inorganic nature is made for organized nature, and the whole globe for man, as both are made for God, the origin and end of all things.

Science thus comprehends the whole of created things, as a vast harmony, all the parts of which are closely connected together, and presuppose each other.

Considered in this point of view, the earth, and all it

contains, the continents in particular, with the whole of their organized nature, all the forms they present, acquire a new meaning and a new aspect.

It is as the abode of man, and the theatre for the action of human societies; it is as the means of the education of entire humanity, that we shall have to consider them, to appreciate the value of each of the physical characters which distinguish them.

The first glance we throw upon the two-fold domain of nature and of history, is enough to show that the parts performed by the different countries of the globe, in the progress of civilization, present very great differences. The three continents of the South, Australia, Africa, — I except Egypt, which scarcely belongs to it, — and South America, have not seen the birth of either of the great forms of civilization which have exercised an influence on the progress of the race. Down to times very near our own, the scene of history has hardly passed the boundaries of Asia and of Europe. Upon these two continents of the Ancient World, all the interests of the great drama, in which we are at once actors and spectators, is concentrated. Another continent, that of North America, has just been added, and is preparing itself to play a part of the first importance.

In the earliest ages of the world, Asia shines alone. It is at once the cradle of civilization, and that of the nations which are the only representatives of culture, and which are carrying it, in our days, to the extremities of the world. Its gigantic proportions, the almost infinite diversity of its soil, its central situation, would render it suitable to be the continent of *the germs*, and the root of that immense tree which is now bearing such beautiful fruits.

But Asia has yielded to Europe the sceptre of civilization for two thousand years. At the present day, Europe is still unquestionably the first of the civilizing continents. Nowhere on the surface of our planet has the mind of man risen to a sublimer height; nowhere has man known so well how to subdue nature, and to make her the instrument of intelligence. The nations of Europe, to whom we all belong, represent not only the highest intellectual growth which the human race has attained at any epoch, but they rule already over nearly every part of the globe, and are preparing to push their conquests further still. Here, evidently, is the central point, the focus where all the noblest powers of humanity, in a prodigious activity, are concentrating themselves. This part of the world is, then, the first in power, the luminous side of our planet, the full-grown flower of the terrestrial globe.

And yet what a contrast between this moral grandeur and the material greatness of this, the smallest of the continents! Nothing in it strikes us at the first glance. Europe does not astonish us by those vast areas which the neighboring continent of Asia embraces. Its loftiest mountains scarcely reach to half the height of the Himalaya and the Andes. Its plateaus, those of Bavaria and Spain, hardly deserve the name, by the side of those of Tubet and of Mexico. Its peninsulas, what are they in comparison with India and Arabia, each of which forms a world by itself? Its seas, the Mediterranean and its gulfs, are far from having the proportions of the vast ocean which bathes the Asiatic peninsulas. Nowhere those great rivers, those immense streams, that water the boundless plains of Asia and America, and are their pride. Nowhere those virgin forests, which cover immense regions, and make them impenetrable to man; none of those deserts, whose startling and terrible aspect, under other climes, appalls us by their immensity. We see there neither the exuberant fruitfulness of the tropical regions, nor the vast frozen tracts of Siberia; we feel there neither the overwhelming heats of the equator, nor those extremes of cold which annihilate all organic life.

In the productions of organized nature, the same modesty still. The plants, the trees, do not attain to the height and growth which astonish us in the regions of the tropics. Neither the flowers, nor the insects, nor the birds, show that variety and brilliancy of colors, which distinguish the corolla of the flowers, and the plumage of the birds, bathed incessantly in the waves of light of the equatorial sun. All the tints are softened and tempered down.

How reconcile this apparent inferiority with the brilliant part Europe has performed among the other continents? This coincidence between the development of humanity in Europe, and the physical nature of this continent, can it have been only an accident? Or may this part of the world have concealed, under such modest appearances, some real superiorities, which have rendered it more suitable than any other to play so distinguished a part in the history of the world? This is a problem, stated by the great facts I am pointing out, the solution of which we must seek by study.

But a third continent, unknown in the history of ancient days, North America, has also entered the list, and is advancing with giant steps; for it has not to recommence the work of civilization: civilization is transported thither ready made. The old nations of Europe, exhausted by the difficulties of every kind which oppose their march, turn with hope their wearied eyes towards this new world, for them the land of the future. Men of all languages, of every country, are bringing hither the most various elements, and preparing the germs of the richest growth. The simplicity and the grandeur of its forms, the extent of the spaces over which it rules, seem to have prepared it to become the abode of the most vast and powerful association of men that has ever existed on the surface of the globe. The fertility of its soil; its position, in the midst of the oceans, between the extremes of Europe and of Asia, facilitating commerce with these two worlds; the proximity of the rich tropical countries of Central and South America, towards which, as by a natural descent, it is borne by the waters of the majestic Mississippi, and of its thousand tributary streams; all these advantages seem to promise its labor and activity a prosperity without example. It belongs not to man to read in the future the decrees of Providence. But science may attempt to comprehend the purposes of God, as to the destinies of nations, by examining with care the theatre, seemingly arranged by Him for the realization of the new social order, towards which humanity is tending with hope. For the order of nature is a foreshadowing of that which is to be.

Such, gentlemen, are the great problems our study

lays before us. We shall endeavor to solve them by studying, first, the characteristic forms of the continents, the influence of these forms on the physical life of the globe; then, the historical development of humanity. We shall have succeeded, if we may have shown to you,

1. That the forms, the arrangement, and the distribution, of the terrestrial masses on the surface of the globe, accidental in appearance, yet reveal a plan which we are enabled to understand by the evolutions of history.

2. That the continents are made for human societies, as the body is made for the soul.

3. That each of the northern or historical continents is peculiarly adapted, by its nature, to perform a special part corresponding to the wants of humanity in one of the great phases of its history.

Thus, nature and history, the earth and man, stand in the closest relations to each other, and form only one grand harmony.

Gentlemen, I may treat this beautiful subject inadequately; but I have a deep conviction that it is worthy to occupy your leisure, as it will occupy for a long time to come, if I am not mistaken, the most exalted minds, and those most ripened for elevated researches. For him who can embrace with a glance the great harmonies of nature and of history, there is here the most admirable plan to study; there are the past and future destinies of the nations to decipher, traced in ineffaceable characters by the finger of Him who governs the world. Admirable order of the Supreme Intelligence and Goodness, which has arranged all for the great purpose of the education of man, and the realization of the plans of Mercy for his sake!

Be pleased always to remember, in my favor more than for yourselves, that the path of science is often difficult and beset with rugged cliffs. The traveller doubtless gathers many flowers on the way. But the tree of Science, which bears the noblest fruits, is placed high up on precipitous rocks. It holds out to our view these precious fruits from afar. Happy he who by his efforts may pluck one of them, even were it the humblest. He values it, then, by what it has cost him. I have made the attempt, and this fruit I offer to you. In default of beauty, may you find therein the savor that I have tasted myself.

After what we have just said of the importance of the geographical forms of the crust of the globe, you will not be surprised, gentlemen, that these very forms of contour and relief, although so far entirely outside, and the arrangement of the great terrestrial masses, are to be the first subject to occupy our attention.

Each of these masses is a solid, of which we are not able to ascertain the configuration, except by considering it at once in its horizontal dimensions and in its vertical dimensions; that is, in its extent and in its contours; then in the varieties of relief which its surface presents. It is in this twofold point of view, and that of their relative situation, that we must first of all study them.

The contours of the continents, as they are shown by the maps before your eyes, are nothing else than the delineation of the line of contact between the lands and the horizontal surface of the oceans. This line is a true curve of level, the sinuosities of which depend entirely upon the plastic forms of the continent itself. It would change its form completely by the relative depression or elevation of the seas. Such as it is, it presents us an almost infinite variety of bends, in and out, which at the first glance seem perfectly irregular and accidental. Yet a more attentive study, and a comparison of the characteristic figures of the continents, enable us to perceive certain features of resemblance and a general disposition of their parts, which seem to indicate, as we shall see, by-and-by, the existence of a common law which must have presided over their formation.

These grand analogies, and these characteristic differences of form and grouping, simple and evident as they appear to us when they have once been pointed out to our attention, have nevertheless been discovered only by degrees, and in succession, by the most eminent minds.

Lord Bacon, the restorer of the physical sciences, first opened the way by remarking that the southern extremities of the two worlds terminate in a point, turned towards the Southern Ocean, while they go on widening towards the north.

After him, Reinhold Forster, the ~~learned and judicious~~ companion of Captain Cook in his second voyage round the world, took up this observation and developed it to a much greater extent. He points out substantially three analogies, three coincidences in the structure of the continents.

The first is that the southern points of all the continents are high and rocky, and seem to be the extremities of mountain belts, which come from far in the interior, and break off abruptly, without transition, at the shore of the ocean. Thus America, which terminates in the rocky precipices of Cape Horn, the last representatives of the already broken chain of the Andes; thus Africa, at the Cape of Good Hope, with its high plateaus and its Table Mountain, which rises from the bosom of the ocean to a height of more than 4,000 feet; thus Asia with the peninsula of the Deccan, which sends out the chain of the Ghauts to form the high rocks of Cape Comorin; Australia, lastly, whose southern extremity presents, at Cape Southeast, of Van Diemen's Land, the same abrupt and massive nature.

A second analogy is, that the continents have, east of the southern points, a large island, or a group of islands more or less considerable. America has the Falkland Islands; Africa, Madagascar and the volcanic islands which surround it; Asia has Ceylon; and Australia, the two great islands of New Zealand.

A third peculiarity of configuration, common to these same parts of the world, is a deep bend of their western side towards the interior of the continent. On this side their flanks are as if hollowed into a vast gulf. In America, the concave summit of this inflection is indicated by the position of Arica, at the foot of the high Cordillera of Bolivia. In Africa, the Gulf of Guinea expresses more strongly still this characteristic feature. It is more feebly marked in Asia by the Gulf of Cambaye, and the Indo-Persian Sea; it reappears fully in Australia, where the Gulf of Nuyts occupies almost the whole southern side.

Forster did not stop here. Seeking to explain to himself these remarkable coincidences in the structure of the great terrestrial masses, he arrived at the conclusion that they were due to a single cause, and that this cause was a great cataclysm coming from the south-west. The waters of the ocean, dashing violently against the barrier the continents opposed to them, ground away their sides with fury, scooped out the deep gulf open towards the south-west, swept off all the movable earth from the southern side, and left nothing standing but those rocky points, that formed only the skeleton. The islands on the east would be only the accumulated ruins of this great catastrophe, or the pieces of the continent protected from total destruction by the jutting point which received the first shock.

This hypothesis, bold as it is ingenious, was admitted by several of the most distinguished contemporaries of Forster. Pallas, among others, the celebrated northern traveller, inclines to receive this general cataclysm from the south-west, which seems to him to explain the great geological phenomena he had observed in the north of Asia. He attributes to it the hollowing out of the deep gulfs which cut into the south of Europe and of Asia, and the formation of the great plains on the north, of those of Siberia, in particular.

The whole ground, according to him, would be composed of earth torn from the southern countries, transported by the waves of the ocean, and by them deposited in these places, after their fury had been spent upon the Himalaya, or the great table land of Asia. It is thus that he explains the presence in Siberia of fossil elephants and of mammoths, and a multitude of other animals and plants which live at the present day only under the sky of the tropics. He remarked, moreover, in support of this hypothesis, that the disproportion existing between the extent of the part of Asia situated south of the Himalaya, compared with that of the vast plains which flank the north of the central mass of the continent, seems to indicate that a great portion of these southern regions has been carried away by this great flood. Pallas, lastly, applies the same observation to America, the western part of which is reduced to a narrow strip, while the region east of the Andes makes almost the whole of the continent.

Seductive as this idea is at the first glance, it is scarcely necessary to say that all that modern geology has taught upon the structure of the mountains, their rise, and the composition of the soil, forbid us to adopt it. It dates from a period when the mind, struck for the first time with the revolutions of the globe, of which it saw the traces everywhere, found no force sufficiently powerful to bring them about, and when water, in particular, seemed the only agent that could be resorted to for their explanation. Nevertheless, it has the merit of binding together, and of fixing, in a precise manner, certain great facts, the existence of which is incontestable.

At a later period Humboldt also shows that he is watching those general phenomena of the configuration of the continents, seemingly destined to reveal the secret of their formation. He first calls our attention to the singular parallelism existing between the two sides of the Atlantic. The salient angles of the one correspond to the reëntering angles of the other; Cape St. Roque in America, to the Gulf of Guinea; the head-land of Africa, of which Cape Verd is the extreme point, to the Gulf of Mexico, so that this ocean takes the form of a great valley, like those the mountainous countries present in such numbers.

Steffens pushed the study of these analogies of the structure of the continents further still, and the picture which he gives us of them opens several new views upon the subject. He remarks, first, that the lands expand and come together towards the north, while they separate and narrow down to points in the south. Now this tendency is marked, not only in the principal masses of the continents, but also in all the important peninsulas which detach themselves from it. Greenland, California, Florida, in America; Scandinavia, Spain, Italy, and Greece, in Europe; the two Indies, Corea, Kamtschatka, in Asia, all have their points turned towards the south.

Passing to the grouping of the continents among themselves, this learned man brings to our view the fact that these great terrestrial masses are grouped two by two, in three double worlds, of which the two component parts are united together by an isthmus, or by a chain of islands; moreover, on one side of the isthmus is found an archipelago, on the opposite side a peninsula.

The purest type of this grouping of the continents is America. Its two halves, North America and South America, are nearly equal in size, and similar in form; they form, so to speak, an equilibrium. The isthmus which unites them is long and narrow. The archipelago on the east, that of he Antilles, is considerable; the peninsula on the west, California, without being greatly extended, is clearly outlined.

The two other double worlds are less regular, less symmetrical. First, the component continents are of unequal size; then the two northern continents are united, and, as it were, joined back to back. Steffens divides them by a line passing through the Caucasus, and coming out upon the Persian Gulf. He thus recombines with Europe a part of Western Asia and Arabia, and gives Africa for its corresponding continent. They are united by the Isthmus of Suez, the shortest and most northern of all. The peninsula found on the east is Arabia, which is of considerable size; the archipelago on the west is that of Greece, which is comparatively of small importance.

This relation is ~~evidently, gentlemen, as you will agree,~~ a forced one. But it seems to me that it would be easy to reëstablish the analogy, so far as the irregularity of structure in the European continent permits, by considering Italy and Sicily, which almost touch Africa by Cape Bon, as the true isthmus. The archipelago is then found on the east, according to the rule, and the peninsula, Spain, on the west.

The third double world, Asia-Australia, is more normal; it approaches nearer the type. The isthmus which unites them is broken, it is true. But that long, continuous chain of islands, stretching without deviation from the peninsula of Malacca, by Sumatra, Java, and the other islands of the Sonde, to New Holland, offers so striking an analogy and parallelism to the isthmus which unites the two Americas, that, before Steffens, Ebel and Lamark had already pointed it out. The great archipelago of Borneo, Celebes, and of the Moluccas, corresponds to that of the Antilles; the peninsula of India, to California.

Here the disproportion between the two continents, as to their extent, is pushed to the extreme. Asia-Australia presents the union of the greatest and the smallest of the terrestrial masses.

These three double worlds exhaust the possible combinations of relations between their component continents. In America, that of the north and that of the south are equal in form and in power; there is a symmetry. In the two others they are unequal. In Europe-Africa, the northern continent is the smallest. In Asia-Australia, it is the continent of the south.

These views of Steffens, even without being justified by a physical theory of the phenomena, are not the less of high interest, and lead us to consider the grouping of the continents under a point of view of the application of which we shall by-and-by see the utility.

But none of the authors who occupied themselves with these questions of configuration and of grouping of the terrestrial spaces, has done so in a manner more happy, and more fruitful in important results, than Carl Ritter.

This founder of historical geography, in the high sense that should be attached to the word, this learned scholar, who has exalted geography to the rank of a philosophical science by the spirit he has breathed into it, applied himself chiefly to investigating what are the fundamental conditions of the form of the surface of the globe most favorable to the progress of man and of human societies. This novel point of view led him to the discovery of relations unperceived until then. We proceed to take cognizance of the principal of them, but at present in a manner wholly external. The signification of these groupings and of these forms will become manifest in the course of our studies.

Ritter showed not only that the lands are more numerous in the northern portion of the earth than in the southern, but that, if we draw a great circle at once through the coast of Peru and the south of Asia, the

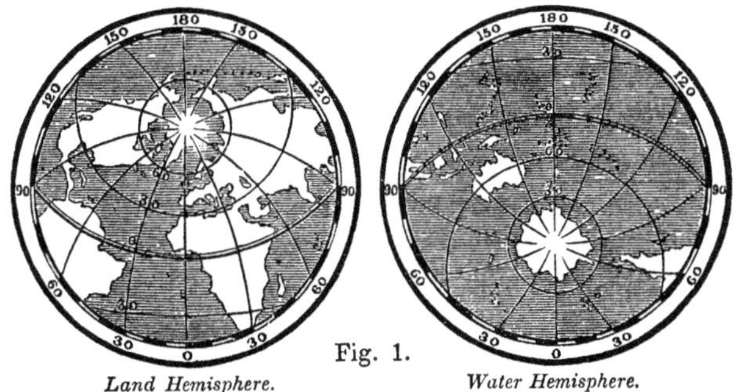

Fig. 1.

Land Hemisphere. *Water Hemisphere.*

surface of the globe is found to be divided into two hemispheres, the one containing the most extensive terrestrial masses, those nearest together and most important; while we behold, in the other, only vast oceans, in which float here and there the peninsular extremities of the principal lands, narrowed and dispersed, and Australia, the smallest and most isolated of the continents. One is then the Continental or Land hemisphere, the other is the Oceanic or Water hemisphere. (See Fig. 1.)

The second general fact, with regard to the grouping of the lands, is that of their combination in two great masses, the Old World and the New World, the forms and structure of which make a striking contrast, and give them a marked character of originality.

Thus the direction of their greatest extension is the inverse in the two worlds. The principal mass of the Ancient World, Asia-Europe, stretches from east to west over one half of the circumference of the globe; while its width is vastly less, and occupies, even in Asia, only a part of the space which separates the equator from the pole. In Europe it is not equal to the sixth part of the earth's circumference. In America, on the contrary, the greatest length extends from the north to the south. It embraces more than one third of the circumference of the globe, and its width, which is very variable, never exceeds a fifth of this amount.

The most remarkable consequence of this arrangement, is, that Asia-Europe extends through similar climatic zones, while America traverses nearly all the climatic zones of the earth, and presents in this relation a much greater variety of phenomena.

The most important of these geographical relations of configuration, that which Ritter was the first to bring prominently forward, and the whole value of which he has explained with rare felicity, is the difference existing between the different continents with regard to the extension of the line of their contours. Some are deeply indented, furnished with peninsulas, gulfs, inland seas, which give to the line of their coasts a great length. Others present a mass more compact, more undivided; their trunk is, as it were, deprived of members, and the line of the coasts, simple and without numerous inflections, is comparatively much shorter.

Considered under this aspect, the three principal continents of the Old World form a remarkable contrast.

Africa is far the most simple in its forms. Its mass, nearly round or ellipsoidal, is concentrated upon itself. It thrusts into the ocean no important peninsula, nor anywhere lets into its bosom the waters of the sea. It seems to close itself against every influence from without. Thus the extension of the line of its coasts is only 14,000 geographical miles, of 60 to the degree, for a surface of 8,720,000 square miles; so that Africa has only one mile of coast for 623 miles of surface.

Asia, although bathed on three sides only by the ocean, is rich, especially on its eastern and southern coasts, in large peninsulas, as Arabia and the two Indies, Corea, Kamtschatka. Whole countries push out into the ocean, as Mandchouria and China. Nevertheless, the extent of this continent is such, that, in spite of the depth of the indentations, there yet remains at its centre a greatly preponderating mass of undivided land, which commands the maritime regions as the body commands the limbs. Asia is indebted to this configuration for a line of coast of 30,800 miles; it is double that of Africa, which is, nevertheless, only one third smaller. Asia, therefore, possesses a mile of coast to 459 square miles of surface.

Of all the continents, Europe is the one whose forms

of contour are most varied. Its principal mass is deeply cut in all parts by the ocean and by inland seas, and seems almost on the point of resolving itself into peninsulas. These peninsulas themselves, as Greece, Scandinavia, repeat to infinity the phenomena of articulation and indentation of coasts, which are characteristic of the entire continent. The inland seas and the portions of the ocean its outer limits enclose, form nearly half of its surface. The line of its shores is thus carried to the extent of 17,200 miles, an enormous proportion compared with its small size; for it is 3,200 miles more than Africa, which is nevertheless three times greater. Europe enjoys one mile of coast for every 156 square miles of surface. Thus it is the continent most open to the sea, for foreign connections, at the same time that it is the most individualized, and the richest in local and independent districts.

In this regard there is, as we see, a sensible gradation between the three principal continents of the Old World. Africa is the most simple; it is a body without members, a tree without branches. Asia is a mighty trunk, the numerous members of which, however, make only a fifth of its mass. In Europe, the members overrule the principal body, the branches cover the trunk; the peninsulas form almost a third of its entire surface. Africa is closed to the ocean; Asia opens only its margins; Europe surrenders to it entirely, and is the most accessible of all the continents.

America repeats the same contrasts, although in a less decided manner. North America, like Europe, is more indented than South America, the configuration of which, in the exterior at least, reminds us of the forms of Africa, and the uniformity of its contours. The two continents of the New World are more alike. Nevertheless, the line of the shores is much more extended in North than in South America. It is 24,000 miles in the former, or one mile of coast to 228 square miles of surface; in the latter, it is 13,600 miles, or a mile of coast for 376 miles of surface.

The following table represents these differences of configuration of the continents by numerical proportions. The mile here employed is the geographical mile, of 60 to the degree. It is the only one we shall make use of in the course of these lectures.

Countries.	Surface in square miles of 60 to a degree.	Length of line of shores.	Square miles for 1 mile of coast.
Europe, ...	2,688,000	17,200	156
Asia, ...	14,128,000	30,800	459
Africa, ...	8,720,000	14,000	623
Australia, ...	2,208,000	7,600	290
North America, . .	5,472,000	24,000	228
South America, . .	5,136,000	13,600	376

It is to Ritter, moreover, as well as to Humboldt, that science is indebted for the appreciation of the value of the relations of size, of relative position of each of the continents, the influence of which, in nature and in history, will appear constantly greater the further we advance in our inquiries.

The exposition we have just made has shown us at once differences and analogies in the forms and dis position of the continental or land masses. The differ ences prove that each continent, or each group of continents, has a character peculiar to itself, and in some sort individual. The analogies lead us to suspect the existence of a general law; they disclose an arrangement which cannot be without a purpose; now, this purpose it will be our duty to seek to comprehend, if we would attain to the true understanding of this part of Creation.

LECTURE II.

Recapitulation — Vertical dimensions or forms of relief — Difficulties presented by their study — Usefulness of profiles — Great influence of differences of height — Elevations in mass, and linear elevations — Importance of the former — Labors of Humboldt and Ritter on this subject — Examination of the general features of relief of the continents — A great common law embracing them all.

LADIES AND GENTLEMEN: —

The conclusion of the preceding lecture was devoted to a simple examination of the most prominent forms which the continents present to us, and such as the line of contact of the lands with the horizontal surface of the oceans exhibits to the eye. In this first review, we have followed, step by step, in their discoveries, the men of science who were the first to point them out. We have recognized, with Lord Bacon and Forster, the tapering form of the southern points of the continents, their gulfs on the west, and their islands on the east; with Pallas, the situation of the great plains in the north of the Old World, and the east of the New; with Humboldt, the winding forms and parallel shores of the great oceanic valley bearing the name of the Atlantic; with Steffens, the enlargement of all the lands towards the north, and the characteristic grouping of the continents in three double worlds. Ritter, finally, has shown us how almost all the lands are combined in one hemisphere, which may be contrasted as a continental hemisphere with the other, which is almost entirely covered with water; how the lands, in their turn, are grouped in two principal masses, in two worlds, the Old and the New, differing in extent, in forms of contour, in structure, and in direction. This learned man, above all, teaches us to consider the forms of the continents in a light entirely new, by drawing our attention to one of the most characteristic features, and, as we shall see, one of the most important, which had escaped all the geographers before him; and that is the considerable difference the various continents present with regard to the greater or smaller number of indentations of their coasts, and of the lesser or greater extent of the line of their shores, of their more or less gradual contact with the waters of the seas and of the oceans.

All these characteristic differences, their gradation, and, above all, the numerous analogies the forms and the grouping of the great terrestrial masses present, have appeared to disclose a symmetrical arrangement, and, as it were, an organization of the continental masses, owing, doubtless, to a physical law, none the less real on account of its being as yet unknown to science.

We shall have occasion hereafter to estimate the value and the influence of these relations, which we have merely stated. But to complete our preparatory study, it is not enough to have taken cognizance of these horizontal forms. We must further make ourselves acquainted with the vertical configuration of the surface, also, of the continents bathed by the atmosphere; that is to say, it is necessary to grasp the most essential features of their relief, so intimately combined with the varieties of their horizontal forms, and moulding in such various manners the different countries of the globe. It is only after having considered the continents under this second point of view, that we shall have the elements necessary to understand the great phenomena of the life of the globe.

But a great difficulty of this study is, that the eye cannot distinguish the elevations on the maps, as well as the contours; besides, physical maps are still wanting for a great part of the earth, and have only been made by nations the most advanced in civilization. In this regard, America deserves to have one of her own, and every friend of science should lend the aid of his good wishes to the accomplishment of so desirable a result.

To remedy these difficulties, we must avail ourselves of profiles. You will easily understand what a profile is, by casting a glance at the plaster model before you, representing one of the most rugged and broken parts of the Swiss Jura. If, cutting it perpendicularly in its length, you place yourselves in front of the section, the line formed by the edge of the surface will present you a profile of the kind which we shall make use of. It would be of the highest interest to preserve, in these sections, the true proportion between the heights and the horizontal extent, such as it is in nature; but it is not possible to do so without making use of drawings on a very large scale, and when it is intended to represent only a very small portion of the terrestrial surface. If, however, we were to make transverse sections of an entire continent, the extent of the horizontal dimensions.

compared with that of the vertical dimensions, would be so great that the latter would become imperceptible to the eye. We are, therefore, obliged to enlarge the scale of heights a certain number of times, in order to render them sufficiently distinct; and this has been done in the numerous profiles that are here exhibited. (See plates II. and III.) You will understand the necessity of this disproportion, if you consider the fact that the loftiest mountain of the globe is only six miles high, while the diameter of the earth is nearly eight thousand miles; so that, representing it in its true proportions, the Dhavalagiri, with its 28,000 feet, would be raised only a twelfth of an inch on a globe ten feet in diameter.

Nevertheless, it would be a mistake to draw from this fact the conclusion that the knowledge and study of these forms, so insignificant in appearance, have but a slight importance. This element, on the contrary, is so essential, that an elevation of level of 350 feet, for example, which is only that of many of our public edifices, is sufficient to diminish the mean temperature of a place by one degree of Fahrenheit; that is to say, the effect is the same as if the place were situated sixty miles further ~~south~~. A few thousand feet of height, which are nothing to the mass of the globe, change entirely the aspect and the character of a country. The excellent vineyards bordering the banks of the Swiss lakes become impossible at 1,000 feet, at 500 even, above their present level; and the tillage, the occupations of the inhabitants, take here a quite different character. A thousand feet higher still, and the rigor of the climate no longer permits the fruit trees to flourish; the pastures are the only wealth of the mountaineer, for whom industry ceases to be a resource. Higher still, vegetation disappears, with it the animals, and soon, instead of the smiling pictures of the plain and the lower valleys, succeeds the spectacle of the majestic but desolated regions of eternal ice and snow, where the sound and animation of life give place to the silence of death.

In truth, all the life of the globe is spread on the surface, and the whole space comprised between the bed of the oceans and the regions of the atmosphere, habitable for organized beings, forms only a thin pellicle round the enormous mass of our planet.

The physical position of a place, as I would call its altitude, or its elevation in the atmosphere above the level of the seas, is, then, the necessary complement of its geographical position. In considering only places situated in a region of small extent, this element is even far the most important to know.

Although the forms of relief are infinitely varied, it seems to me that we may refer them to two great classes, admitting of numerous modifications.

1. The elevations in mass, and by great surfaces, which are called *plains*, or *lowlands*, when they are elevated only a little above the level of the oceans, and *plateaus*, or *table lands*, when their elevation is more considerable, and presents a solid platform, a basis of great thickness.

2. The linear elevations and the chains of mountains, which are distributed over the surface and on the borders of the plains and of the table lands, or more rarely scattered in isolated groups.

Of these three forms, those which strike us most at the first glance, are the mountains; and so the geographers have occupied themselves with these in the first place. Buache, of the French Academy, in the middle of the last century, was the first who attempted to comprise in a systematic order the whole of the mountains of the earth; but he was too often obliged to supply by imagination the want of positive knowledge; and I mention here his essay on the connection of the mountains of the globe, only to point out the first step that was taken in this path. After him, Buffon made the important observation, that the principal mountain chains of the Old World follow the direction of the parallels, and those of the New World, the direction of the meridians; and that the secondary chains follow the inverse in both.

This predilection for the mountains lasted a long time; we may say that it still prevails in geology. Although the upheaval of the great surfaces, horizontal or slightly inclined, the elevation of entire continents, may be perhaps a more essential fact in the physical history of our globe, than that of a chain of mountains; nevertheless, geology has scarcely occupied itself except with the latter, and seems almost ready to admit that the upheaval of mountain chains is the principal fact, and that of the large surfaces and of the plateaus the accompaniment. This is not the place to discuss this great question, but we are bound to say, that, at all events, in physical geography we cannot be of this opinion.

Although the word *plateau* was introduced into science by Buache, the importance of these elevations in mass in physical geography was not recognized in reality before the time of Alexander Humboldt. He was the first to bring prominently out, by his barometrical sections, the remarkable forms of the plateau of Mexico, and of the high valleys of the Andes. No one of the great physical consequences connected with this structure escaped his penetrating sagacity. After him it was not allowed to neglect the important element of the altitudes, and this great truth remained an acquisition to science.

Carl Ritter soon after applied these principles to the study of all the continents. Drawing from the treasures of his vast erudition, he availed himself of all the documents scattered over thousands of volumes, to give us a true image of the structure of the continents. He distinguished with greater precision the high plateaus of Central Asia and of Western Asia from the low lands which surround them; he exhibited the contrast between the high lands of Southern Africa, and the low plains of the Nile and of Sahara. Each of the countries of the Old World under this new light appeared to our eyes for the first time in its true form, as those of the New World had been revealed by Humboldt.

For a long time still we shall have to persevere in this path which genius has opened, in order to complete by observation the work so happily begun. But have we not another step to take? Shall we not find here, in the midst of this infinite variety of forms of relief, some of those grand analogies which have struck us in the study of the horizontal forms, some of those general facts which authorize us to admit for the elevations, also, some great common law around which the particular facts arrange themselves?

We shall endeavor to solve this important question,

not, gentlemen, by any hypothesis, but by the combination and exposition of the facts recognized in science. For this purpose I may often be obliged to quote figures; but even figures have their eloquence. These, for greater convenience, I shall express in round numbers, as it will sufficiently answer the end I propose.

The examination of the general reliefs of the great masses of dry land on the surface of the globe, leads us, in fact, to the recognition of certain great analogies, certain great laws of relief, which apply, whether to certain groups of continents, or to all the continents taken together, or to the whole earth. I shall point out, one after another, these general facts, supporting them by examples; and, with the aid of the profiles you have before you, I hope to make clear to you the general law which appears to me to follow from them.

1. All the continents rise gradually from the shores of the seas towards the interior, to a line of highest elevation of the masses, and of the peaks surmounting them, to a maximum of swell.

This fact appears trivial in the stating, because it seems so much according to the nature of things. But it is not so for him who knows the geological history of our continents and the revolutions their surface has undergone. The question is asked, why we should not have, in the interior of vast continents like Asia or America, some great depression, the bed of which should be sunk below the surface of the oceans. And in fact this circumstance is not absolutely wanting to our continents; we may cite, as a case of the kind, the great hollow, the bottom of which is occupied by the Caspian Sea. It is known that the surface of this sea, and even of a great part of the surrounding countries, is below the common level of the oceans; further, its basis presents in its southern parts considerable depths. The valley of the Jordan and Dead Sea, together with its lakes and the river, is almost entirely below the level of the Mediterranean. The recent measurements of Bertou, of Russegger, and of several others, among whom I will mention, as the most recent of these bold explorers, an American, Lieut. Lynch, have proved that the level of the Dead Sea is about 1,300 feet below the level of the oceans, and that its depth descends at least as much more. What masks these depressions, moreover, is the water filling them, the surface of which must be considered as forming a part of that of the continents. Besides the three largest of the lakes of Canada, several of the lakes of the Italian Alps, the bed of which sinks below the level of the sea, would appear as similar excavations. We may say the same of the midland seas bordering the European continent on the north and on the south.

2. In all the continents, the line of greatest elevation in the summit of ascent is placed out of the centre, on one of the sides, at an unequal distance from the shores of the seas. From this fact result two slopes, unequal in length and in inclination. This is analogous to what, in mountains, is called the slope and the counter slope.

The length of these two inclined planes estimated approximately and in round numbers, is nearly as follows, in the different transverse sections of the continents represented by the profiles which we have before us (See plates II. and III.)

The first column indicates the length, in geographical miles, of the long slope; the second, that of the short one.

OLD WORLD.—*North to South.*

	Length in Miles.	
	Northern Slope.	Southern Slope.
1.—Eastern Asia. The section begins at the Frozen Ocean, at the mouth of the Jenisei, and terminates in the plains of the Ganges. The culminating region is that of the table lands of Tubet and of the Dhavalagiri, which divides this line into two slopes, ...	2,600	400
2.—Western Asia. From Lake Aral and the plains of the Caspian Sea to the Persian Gulf; culminating point, the coast chain of the Persian Gulf, . . .	900	100
3.—Western Asia. From the plains of Georgia to those of the Euphrates; culminating point, the high chains of Kurdistan, ...	260	80
4.—Asia Minor. From the northern to the southern coast, nearly on the meridian of Cyprus; culminating point, the chain of Taurus, ...	300	50
5 —Central Europe. From the shores of the Baltic to the plains of Lombardy; culminating point, the Tyrolian Alps, ...	450	100
6.—Africa. From the mouth of the Nile to the Cape of Good Hope; culminating point, probably the high plateaus between the sources of the Zambeze and of the Orange river, . .	3,300	600

Plate II. OLD WORLD. — NORTH TO SOUTH.

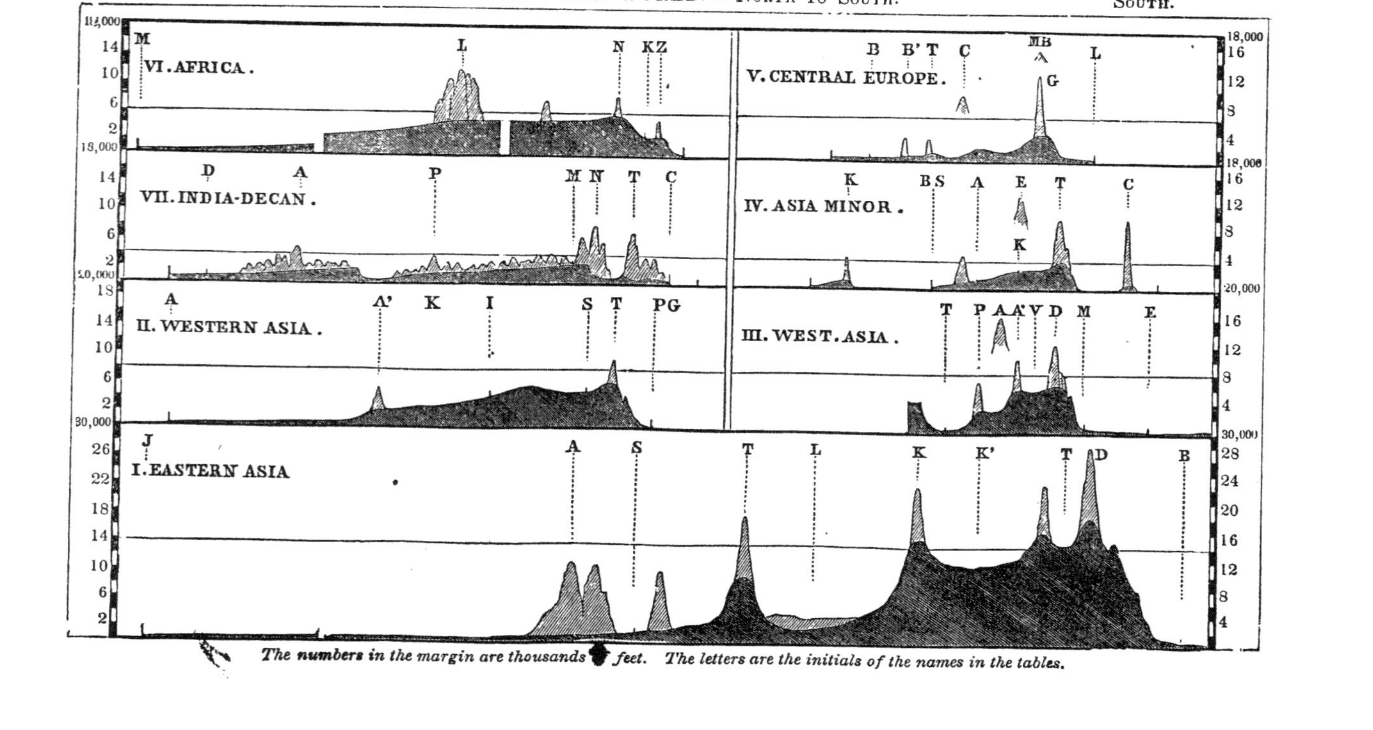

The numbers in the margin are thousands of feet. The letters are the initials of the names in the tables.

Plate III. NEW WORLD. — East to West.

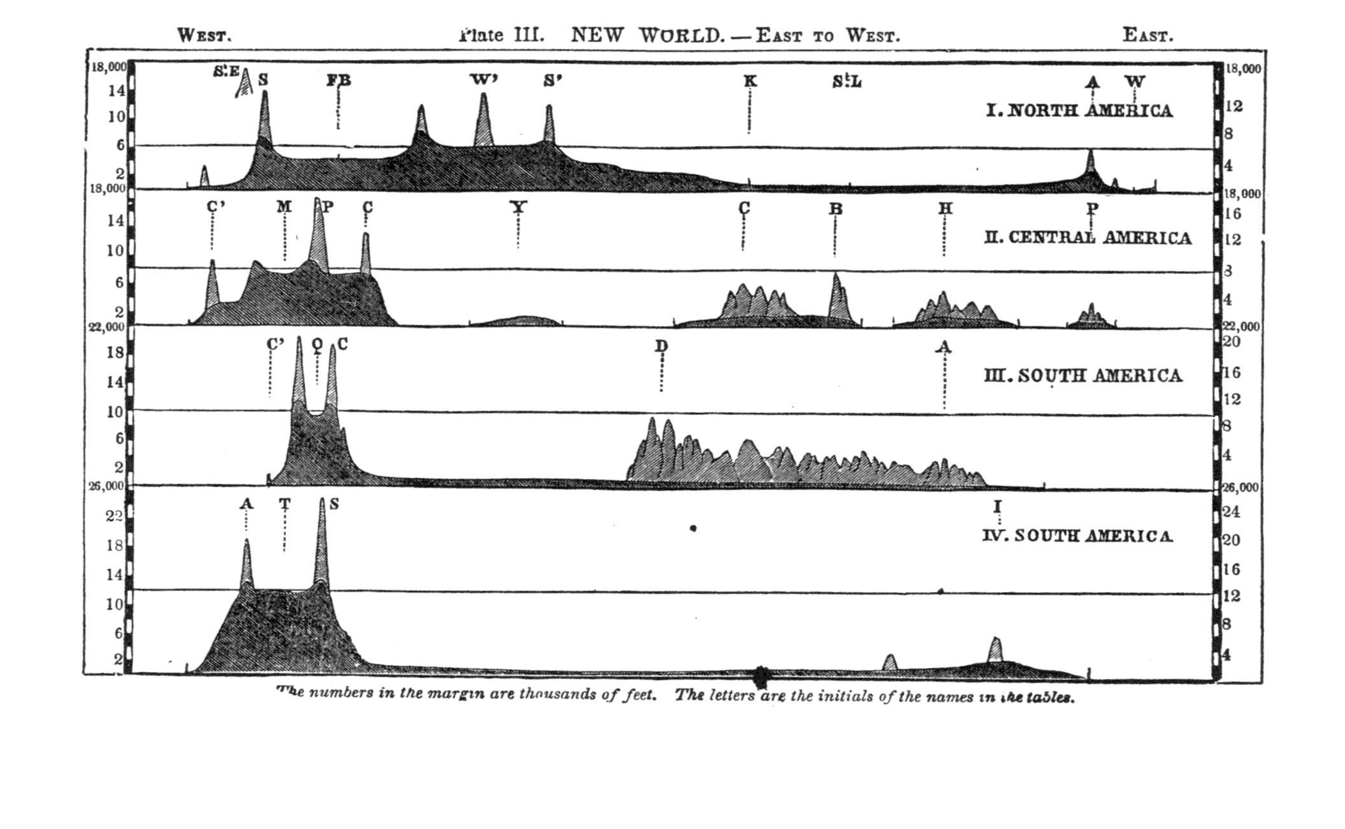

The numbers in the margin are thousands of feet. The letters are the initials of the names in the tables.

NEW WORLD.

	Length in Miles.	
	Eastern Slope.	Western Slope.
1. — North America.		
From Washington to the Bay of St. Francisco; culminating point, the central chain of the Rocky Mountains, ...	1,600	800
2. — Central America.		
From Porto Rico, through Mexico, to the Pacific Ocean — the line slightly broken to take in the Great Antilles; culminating point, the plateau of Mexico, ...	2,000	300
3. — South America.		
From the mouth of the Amazon, through the table land of Peru, to the Pacific Ocean; culminating point, the Chimboraço, . . .	1,850	70
4. — South America.		
From the coasts of Brazil, north to Rio Janeiro, through the Lake of Titicaca, to the Pacific; culminating point, the Nevado de Sorata, ...	1,600	200

We see, by this table, that one of the general slopes of the continents is always, if we take the mean, at least four or five times as large as the other.

3. This law of increase of reliefs is common to the mass elevations, and to the linear elevations; that is, the height of the low lands and of the table lands increases at the same time with the absolute elevation of the mountains. There is a proportional gradation.

This law is exhibited by the following table, containing the principal elements I have used in constructing the profiles. As they are intended to set in a clear light the most general features of the relief of the continents, these profiles do not always follow an exactly straight line, but they sometimes embrace rather a transverse zone. For the same reason I merely indicate, without taking them into the view, several lofty volcanic peaks, isolated like the Ararat, the Erdshish of Asia Minor, which, considered in relation to the general relief of the countries where they are found, are but accidents, and cause only a local modification. They are marked in the tables by an asterisk.

The first column contains the height of the plateaus, the second that of the loftiest corresponding peaks, both in English feet. In plates II. and III. they are both indicated by their initials.

OLD WORLD.—*North to South.*

1.—EASTERN ASIA.	Low l'ds and Table lands.	Highest Mountains.
Coast of the Frozen Ocean, mouth of the Jenisei,	0	0
Plains of Siberia, Barnaul, foot of the Altai Mountains,	380	
Altai, Bjelucha,	....	11,000
Dzoungary, Lake Saisan,	1,300	
Thian-Shan Chain, Bogdo-oola,	....	18,000
Eastern Turkestan, Desert of Lop,	2,000?	
Northern Basis of the Kuenlun,	5,000?	
Chain of Kuenlun,	....	21,000?
Katschi, plateau,	11,000	
Plateau of Tubet,	14,000	
Dhavalagiri,	....	28,070
Benares, plains of the Ganges,	250	
2.—WESTERN ASIA.		
Lake Aral, plains of the Caspian Sea,	0	0
Foot of the table land,	500?	

	Low l'ds and Table lands.	Highest Mountains.
2.—WESTERN ASIA.		
Atak chain, Kabushan, ...	. . . .	4,500
Table land of Khorassan, ...	2,000	
Ispahan, ...	4,400	
Shiraz, ...	4,500	
Tengi-Firesen, coast chain, ...	. . . .	9,000
Coast of the Persian Gulf, ...	0	0
3.—WESTERN ASIA.		
Tiflis, plains of Georgia, ...	1,160	
Pambak, chain, ...	. . . .	7,000
Ararat, the plain around, ...	2,800	
*Ararat, mountain, ...	. . . .	17,000
Ala Dagh, ...	. . . .	10,000
Lake Van, ...	5,400	
Djudid Dagh, ...	. . . .	12,000
Mossul, ...	350	
Euphrates, the plains, ...	200?	
4.—ASIA MINOR.		
Krimea, Tchatur Dagh, ...	. . . .	5,000
Black Sea, ...	0	0
Chains along the coast of Asia Minor, . . .	. . . .	5,000
Table land, Amasia, ...	1,200	
Kaisarieh, table land, ...	3,000	
*Erdshish, mountain, ...	. . . .	13,000
Taurus chain, Bulghar Dagh, ...	. . . .	10,000
Coast of Cilicia, ...	0	0
Cyprus, Stavro-Vouno peaks, ...	. . . .	10,000
5.—CENTRAL EUROPE.		
The coast of the Baltic Sea, ...	0	0
Plains of North Germany, Berlin, ...	100	
The Harz, Brocken, ...	. . . .	3,700
Thuringerwald, ...	. . . .	3,200
Carpathian Mountains, Tatra, ...	. . . .	9,500
Table land of Franconia, ...	800	
Table land of Bavaria, foot of the Alps, . . .	2,100	
The Alps, the Glockner, ...	. . . .	12 800
Plains of Lombardy, ...	200	

6. — AFRICA.	Low l'ds and Table lands.	Highest Mountains.
Mediterranean Sea, mouth of the Nile, . . .	0	0
Plains of Cordofan, ...	1,200	
Table lands of the Zambeze, ...	5,000	
Lupata Chain, ...	. . . .	11,000?
Nieuweveld Mountains, ...	. . . .	9,000
Great Karroo, ...	2,700	
Zwarteberge, ...	. . . .	5,000
Coast, ...	0	0

The same law exists in the great peninsulas of Asia, whose basis is a table land, and which are almost small continents, as India and Arabia. In those of Europe, which are smaller and more irregular, the law is not expressed; yet the Taygetus in Greece, the Sierra Nevada in Spain,—I should add Etna in Sicily, were it not an isolated volcano,—belong to the highest summits of the three peninsulas.

7. — INDIA-DECCAN.	Low l'ds and Table lands.	Highest Mountains.
Plains of the Ganges, Delhi, ...	800	
Aravulli Mountains, ...	. . . .	5,000
Table land of Deccan, Punah, ...	1,800	
Mysore, ...	2,700	
Nilgherries, ...	. . . .	8,800
Travancore Mountains, ...	. . . .	7,500
Palamcotta, ...	300	
Cape Comorin, ...	. . . .	500
8. — ARABIA.		
Mouth of the Euphrates, ...	0	0
Inner Arabia, Nedjed, table land, ...	2,000?	
Yemen, Sana, ...	5,000	
Djebel Yafai, ...	. . . .	6,600
Aden-coast, ...	0	0

NEW WORLD.—*East to West.*

	Low l'ds and Table lands.	Highest Mountains.
1.—NORTH AMERICA.		
Washington,	0	0
The Great Valley of Virginia,	300	
Apalachian Mountains,	....	6,000
Mississippi Valley, St. Louis,	500	
Mouth of the Kansas,	800	
Foot of the Rocky Mountains, F. St. Vrain,	5,000	
Spanish Peaks,	....	12,000
Plateau of Inner California,	6,000	
Wind-River Mountains,	....	13,500
Fremont Basin,	4,300	
Sierra Nevada of California, Shastl Peak,	....	14,000
*St. Elias Peak,	....	17,800
Nueva-Helvetia,	40	
Pacific Ocean,	0	0
2.—CENTRAL AMERICA.		
Porto Rico,	....	3,500?
Haïti,	....	5,000?
Cuba,	....	7,000
Jamaica, Blue Mountains,	....	7,500
Table land of Mexico,	7,500	
Coffre de Perote,	....	13,000
Popocatepetl,	....	19,000
Plateau,	3,000	
Colima Volcano,	....	9,000
Pacific Ocean,	0	0
3.—SOUTH AMERICA.		
Mouth of the Amazon,	0	0
Sierra Parime, Acarai,	....	4,000
Sierra Parime, Duida,	....	8,400
Foot of the Andes,	1,000?	
Volcano of Cayambe,	....	19,500
Quito,	9,000	
Chimboraço,	....	21,300
Pacific Ocean,	0	0

4.—South America.	Low l'ds and Table lands.	Highest Mountains.
Coast of Brazil, ...	0	0
Table land of Brazil, ...	2,000	
Itambe, Sierra de Espinhaço, ...	. . . .	6,000
Sta Cruz de Rio Grande	1,000?	
Nevado de Sorata, ...	. . . .	25,500
Lake of Titicaca, ...	12,800	
Volcano of Arequipa, ...	. . . .	18,300
Coast of the Pacific, ...	0	0

The simultaneous increase of the plateaus and of the higher peaks cannot, then, be doubted, and the preceding figures make all comments unnecessary. The inspection of the profiles permits us also to deduce the following general facts:

4. In the Old World, the long slopes are turned towards the north, and the short slopes towards the south. In the New World, the gentle slopes descend towards the east, and the short and rapid slopes towards the west. Thus, in this respect, each of the two worlds has a law peculiar to itself.

5. In each of the two worlds, the two laws exert their influence. In the Old World, though the principal slope is towards the north, we observe still a gradual decrease of the reliefs from east to west; in the same manner, in the New World, the principal slope is from the west to the east, but it may be seen that the reliefs go on gradually increasing from north to south, as in the Old World. In these two secondary directions of the reliefs, we discover the law of the unequal slopes; in the Old World,

the long slope descends to the west, the short slope to the east; in the New World, the long slope is to the north, the short to the south.

	Low l'ds and Table lands.	Highest Mountains.
1. — Asia - Europe — *from East to West.*		
Dhavalagiri, ...	. . . .	28,000
Table lands of Tubet, ...	14,000	
Hindoo-Kuh, ...	. . . .	20,000
Plateau of Caboul and Kelat, ...	7,000	
Plateau of Persia, ...	4,000	
Caucasus, Elbrus, ...	. . . .	17,800
Asia Minor, plateau, ...	3,000	
Taurus, ...	. . . .	10,000
Alps, Mt. Blanc, ...	. . . .	15,800
Pyrenees, Pic Nethou, ...	. . . .	11,200
Table land of Spain, ...	2,300	
Table land of Bavaria, ...	1,700	
2. — Africa — *from East to West.*		
Abyssinia, Samen Mountains, ...	. . . .	15,000
Plateau of Gondar, ...	7,000	
Plateau of Mandara, ...	4,000?	
Camerun Mountains, ...	. . . .	12,000
Kong Mountains, ...	. . . .	3,000
Table land of the Mandingoes, ...	2,000?	
3. — America — *from North to South.*		
Rocky Mountains, ...	. . . .	13,500
Plateau of California, ...	6,000	
Volcanoes of Mexico, ...	. . . .	17,000
Plateau of Mexico, ...	7,500	
Sta Fe de Bogota, ...	8,500	
Plateau of Quito, ...	9,500	
Chimboraço, ...	. . . .	21,400
Titicaca Lake, ...	12,800	
Sorata, Nevado, ...	. . . .	25,300

6. Generally speaking, the reliefs go on increasing from the poles to the tropical regions. The highest elevations, however, are not placed at the equator, but in the neighborhood of the Tropic of Cancer, in the Old World, (Himalaya, 27° north lat.,) and near the Tropic of Capricorn, in the New World, (Nevado de Sorata, 18° south lat.)

I notice here one of the great compensations, one of the great harmonies, of nature. The effect of this law of arrangement is, to temper the burning heats of these regions, and to give them such a variety of climate as seems not to belong to these countries of the globe. If this order were reversed, and the elevation of the lands went on increasing towards the north, the most civilized half of the globe, at the present day, would be a frozen and uninhabited desert.

7. In fine, a common law combines in a single great fact all we have just said upon the general reliefs of the continent; it may be thus expressed:—

All the long and gentle slopes descend towards the Atlantic and towards the Frozen Ocean, which is only a dependence of it; all the short and rapid slopes, or counter-slopes, are directed towards the Pacific Ocean and towards the Indian Ocean, which is its continuation.

In this point of view, these two great oceans appear as two basins of different geological character.

The Pacific Ocean seems an immense basin which has sunk down, and whose high and ragged edges present on all sides the abrupt terminations of the continents. It is on this great line of fractures, on the borders and all round this ocean, as has been pointed out by Mr. von Buch and other scientific men, that we behold the great majority of the active volcanoes of our globe, arranged like an immense burning crown. If we add to this feature the multitude of volcanic islands scattered over the ocean, we comprehend the idea, expressed by Steffens, that the vast basin occupies the place of a continent of the early ages, uniting the two worlds, but sunk and submerged at present under the deep waters of the ocean, in consequence of one of the latest great revolutions of our globe.

The Atlantic Ocean, on the contrary, would be a simple depression, somewhat in the form of a trough, owing, perhaps, to a lateral pressure, and partly to the tilting motion which lifted up the lands in the neighborhood of the Pacific. Hence, its narrower breadth, the valley form, the absence of numerous islands in the interior of its basin, and the descent of all the neighboring continents by gentle slopes. Nevertheless, if we suppose the lateral force that pressed it in to have been very strong, we may conceive that this valley has a great depth.

Thus, then, gentlemen, a great law, a general law, unites all the various systems of mountains and of plateaus that cover the surface of our globe, and arranges them in a vast and regular system of slopes and counter-slopes. Considered with reference to the present state of geology, this result is astonishing. The study of the relative ages of the different systems of elevations, teaches us that each of them has existed a long time separately. One appeared in one country at a given epoch, another in another. The continents are only formed, so to speak, by piecemeal, in the train of the geological epochs; and, nevertheless, the definitive result makes a whole, composed of parts subordinated to each other in a certain system, which might be called an organism in this order of things.

This is not the moment to press the consequences of so remarkable a fact. It is enough for me to have pointed it out to your attention.

I will add only, that the geological researches place beyond a doubt the existence of an intimate relation between the height of the mountains and of the plateaus, and the epoch of their appearance above the surface of the oceans. The most ancient chains of mountains are the least elevated; while the colossal grandeurs of the Andes and the Himalaya bear the traces of an upheaval comparatively very recent. In America, from the coasts of Brazil to the high table lands of Bolivia, and from the Atlantic shores to the Rocky Mountains; in Europe, from the mountains of Scandinavia to the summit of the Alps, we meet with upheavings successively less ancient. In the two worlds the continental masses have then become greater in the lapse of the ages, not by chance, but in two determinate directions; that is, in a geographical order, from the north to the south of the Old World, and from the east to the west in the New; and I think we may hence infer, that from the moment when the oldest lands we know emerged, the continents have had a tendency to form themselves on the spot where they now are.

We see that here, as elsewhere, all is done with order and measure, and according to a plan which we shall have a right to believe was foreseen and intended, when we shall have studied all the consequences of this arrangement of the continental masses.

LECTURE III.

Distribution of the table lands, the mountains, and the plains in the different continents; the Old World that of plateaus, the New World that of plains — The basin of the oceans; this inquiry completes the study of the plastic forms of the earth's crust — Division and characteristics of the oceans; their contours and their depth — Comparison of the latter with the mean elevation of the continents — Conclusions — Necessity of considering the physiology of the continental forms — Point of view which should be taken — Law of the development of life.

Ladies and Gentlemen: —

In our last lecture we carried our examination into the general forms of relief of the continents. Our investigation has permitted us to establish the existence of a great common law of slopes and counter-slopes, of increase and decrease of reliefs. The entire continents, as well as the mountains, have two principal unequal slopes; the long and gentle slopes descend towards the Atlantic and Frozen Ocean, the short and steep slopes towards the Pacific Ocean.

To finish this subject, it remains to say a word upon the distribution of the table lands, of the mountains, and of the plains in the different continents.

The distribution of these three great forms of relief — is it uniform or not? Or are there not some characteristic differences, to be pointed out in this regard, between the continents? Does not the form of the elevated table lands prevail in one part of the world, the form of the plains in another, the form of the mountains, finally, in yet a third? If we call to mind the important influence each of these forms exerts on the climate, the productions, and on the conditions of existence, and growth of the nations, we shall regard this question as one of those which most concern our subject.

Considered in this point of view, the continents present, in reality, remarkable differences.

The Old World, ~~as we have learned from the study of its reliefs,~~ is that of table lands and mountains. No continent exhibits plateaus so elevated, so numerous, so extensive, as Asia and Africa. Instead of one or two chains of mountains, like the Andes, Central Asia is traversed by four immense chains, supporting vast table lands of from 5,000 to 14,000 feet in elevation, and the loftiest mountains of the globe.

The extent of this elevated region is more than 2,400 miles long, by 1,500 miles broad. The principal mass of Western Asia is nothing but a plateau, from three to six thousand feet in height. Africa, south of Sahara, seems to be only an enormous pile of uplifted lands. It has been calculated that the mountains and plateaus of Asia cover five sevenths of its surface, while the plains occupy only two sevenths. In Africa, the high regions form two thirds of the continent, the plains only one third.

If I call the Old World the world of plateaus, it is not because great plains are wanting there. The whole north of Europe and of Asia is merely a boundless plain, and from the shores of Holland, through Germany, Russia, the Steppes of the Caspian and Siberia, the traveller may cross the Ancient World from the Atlantic to the Pacific Ocean, for a distance of more than six thousand miles, without encountering an eminence of more than a few hundred feet in height. In Africa, also, the plains of Sahara extend 2,500 miles in length, by 1,000 in breadth. But the situation of these plains of the Old World under the frozen sky of the north, and under the fires of the tropics, together with the nature of their soil, takes from them all their importance. The one is a frozen waste, a Siberia; the other a burning desert; and neither the one nor the other is called to play an essential part, nor do they impress upon their respective continents their essential character.

The New World, on the other hand, is the world of plains. They form two thirds of its surface; the plateaus and the mountains, only one third. The high lands form only a narrow band, crowded upon the western coast of the two continents. Almost the whole East runs into immense plains, covering it, one might say, from pole to pole. From the Frozen Ocean to the Gulf of Mexico, over an extent of nearly 2,400 miles, we cross only insignificant heights. From the llanos of the Orinoco to the banks of La Plata, we traverse more than three thousand miles of low plains, slightly interrupted by the somewhat more elevated regions of western Brazil; they are prolonged even to the Pampas of Patagonia, 600 miles further south, to the southern extremity of America. The length of the rich plains watered by the Marañon, in the direction of the current, is nearly 1,600 miles; and what are the plains of the Amazon and the Mississippi, compared with those of Siberia and Sahara? A happy climate, a rich and fertile soil, a wonderful vegetation, prodigious resources — they have all that makes the prosperity of a country, who does not see that here is the character of America, here lies the future of the New World, while the countries of mountains and plateaus seem destined to play only a secondary part?

Even in this regard the two worlds have each its own character, and form a great contrast; so that we may say that, in one of the hemispheres, the plateaus and the mountains predominate, while, in the other, the plains give the important and essential feature of relief.

Finally, if we were seeking for a continent where the form of mountains, without plateaus at their base, should be the characteristic feature, it would be necessary to name Europe, comprehending in it only Western Europe without Russia; that is, historical Europe, the true Europe after all. This continent, with Russia, has three fourths of plains to one fourth of mountainous country; but, leaving out Russia, it is quite otherwise. Traverse Europe from one end to the other, whether over its central mass or its peninsulas, you will find everywhere its soil modified, cut in all directions by chains of mountains intersecting each other. In all this part of the continent, the largest existing plain, that of Northern Germany and Poland, is only six hundred miles long by two hundred broad. It is the extremity of the great Asiatic plains in the north. The other plains, as those of France, of Hungary, of Lombardy, are smaller in extent, and do not deprive this part of the continent of the mountainous character essentially belonging to it.

We have now considered the configuration of the general forms of the continents; let us not forget that this is only one half of the plastic forms of the earth's crust. There is another, which, though hidden from our sight, is none the less entitled to our interest. It is the basin of the oceans.

The positive forms of the lands which we have studied determine negatively for the oceans, whether horizontally or vertically, certain forms no less characteristic. We ought, then, to examine the character of the ocean basins, in the two-fold relation of the forms and the indentations of their shores, and of their depth. But the time presses, and I must be brief; limiting myself to the essential facts, I shall omit all that does not touch directly upon my subject.

The continents determine the general outlines of the great ocean basins. The indentations of their coasts give the articulation of the shores of the oceans; the islands, by their disposition, by their less or greater frequency, furnish what else is wanting to complete their character; the one is the counterpart of the other; they are the same forms, but in an inverse order.

Two great oceans, the Pacific and the Atlantic, corresponding to the two worlds, surround, on almost all sides, the principal terrestrial masses. We may detach from the Pacific, the Indian Ocean, which, though belonging to it, has some special characteristics; and separate from the Atlantic, the Northern Frozen Ocean, the position of which gives it a particular physiognomy. As to the great Southern Sea, we may consider it less as an ocean by itself than as a common reservoir, whence issue, so to speak, all the seas of the globe, to make their way into the lands.

The Pacific, the Indian Ocean, the Atlantic, correspond to the three double worlds which we have distinguished, following Steffens, and separate them from one another. Each of them, also, is divided into a northern and a southern basin, except the Indian Ocean, which, on this account, is only a half ocean.

The general forms of the contours of these three oceans have, as a common feature, a wide opening towards the south, and are narrowed to a point on the north, just the reverse of the continents. Each of these has, meantime, a form peculiar to itself. The Pacific Ocean is an oval, wide open on the south, the sides coming nearer and nearer together towards the north, so as to leave, between America and Asia, only the narrow passage of Behrings' Straits, by which it communicates with the Frozen Ocean.

The Indian Ocean has the form of a triangle, with the vertex turned to the north; the Atlantic that of a valley with nearly parallel sides, which, narrowed for a moment, then broadens into the Frozen Ocean.

The oceans differ, moreover, in the mode of articulation of their shores. These indentations have very various forms, which I will classify, for the moment, under three species: the *gulfs*, like that of Bengal; the *land-locked seas*, isolated from the rest of the ocean by peninsulas and chains of islands, like the Sea of Japan, the Sea of Okhotsk; the *inland seas*, surrounded on all sides by the land, in a continuous manner, like the Mediterranean and the Baltic.

Considered with reference to the indentations, the three oceans have their own respective characters, and we find that in each, one of these three forms predominates.

The Pacific Ocean is that of the land-locked or closed seas; for there are no less than five of considerable size along the coast of Asia: the Sea of Behring, closed in by the peninsula of Aljaska, and the chain of the Aleutian Islands; the Sea of Okhotsk, enclosed by the peninsula of Kamtschatka, and the series of the Kurile Islands; the Sea of Japan, shut in by the island chain of this name; the northern Chinese Sea, locked by the islands of Lieu-Khieu and Formosa; the Southern Sea of China, locked by the Philippines, Borneo, and the peninsula of Indo-China. We may almost call the Vermilion Sea, or Gulf of California, an inland sea, it being the only indentation of this ocean somewhat marked, on the American coast.

The Indian Ocean is that of gulfs; for the two great Gulfs of Bengal and the Persian Sea impress upon it its character. It pushes, besides, into the interior two mid-land seas, the Persian Gulf and the Red Sea, which detach the peninsula of Arabia from the rest of the continent.

The Atlantic Ocean is that of inland seas. No one advances further into the lands, piercing into the very heart of the Old World and the New. There are, at least, four mediterraneans, without taking into the account the Polar seas; two on the European side, the Mediterranean, properly so called, divided into three great basins; the Eastern, the Western, the Black Sea, not to mention several others of small size, and the

Baltic; two on the coasts of the New World, the Gulf of Mexico and Hudson's Bay. Neither is the form of land-locked seas wanting here also; the Northern Ocean, on the coasts of the Old World, the Caribbean Sea in Central America, closed by the peninsula of Yucatan and the chain of the greater and lesser Antilles; the Gulf of St. Lawrence, locked by the peninsula of Nova Scotia and Newfoundland, are the proof. The great gulfs are represented by those of Guinea and of Biscay. The Atlantic Ocean is, then, the most articulated, the most indented of the oceans, and that which, by its blending with the lands, approaches the nearest to the character of the inland seas. It is, if I may venture to say so, the most *maritime* of the oceans, as the Pacific is the most truly *oceanic.*

The islands, finally, are one of the most interesting characteristics of the oceans. There are two species to be distinguished: the *continental* islands, which their proximity, their size, their geographical character, their forming a line with the mountain chains of the firm land, prove to be a dependence of the continents; and the *pelagic,* or oceanic islands, dispersed singly, or in groups, at a distance from the lands, over the vast surface of the ocean, of small dimensions, and always of a volcanic or coralline character.

The Pacific Ocean is far the richest in islands, whether continental or pelagic. The Indian archipelago and that of New Holland is the largest continental archipelago existing on the surface of the globe; and the thousands of pelagic islands with which the centre of this ocean is studded, have nowhere else their parallel.

The Atlantic Ocean possesses still, in the group of the Antilles, the British Isles, and those of the Mediterranean, continental archipelagos of great importance; but the pelagic islands are poorly represented there by the groups of the Azores, Madeira, the Canaries, Cape Verd, St. Helena, and some other small islands lost in the midst of the ocean.

The Indian Ocean is scanty in both. Madagascar and Ceylon, one by a main-land, represent the continental islands. Here and there a few volcanic islands, as Mauritius and Bourbon, represent the pelagic.

Each ocean, therefore, differs from the others, in some peculiarities of character; and we readily conceive how these circumstances may modify their importance, with regard to the facility or the difficulty they may bring into the relations of exchange which commerce establishes among all the nations of the world.

Let us now see what is known of the vertical dimensions, or of the configuration of their basin.

The basin of the oceans is depressed below the face of their waters, as the continents are elevated in the atmosphere above the same surface level. It may, then, be said, that we know not one half of the reliefs of the solid crust of our globe, for more than two thirds are concealed from our observation by the seas that cover them. It would, nevertheless, be of the highest interest for geology, as well as for the physics of the globe, to ascertain the forms, the depth, and the nature of the bottom of the oceans. But though we have numerous soundings, executed in the neighborhood of the shores, to meet the wants of navigation, we have only a very few in the interior, and in the deepest parts of the oceans. These operations require a consumption of time, and an amount of labor, which will always render them rare. Recourse has been had to hypotheses, while waiting for positive information on the subject.

In the neighborhood of the continents, the seas are often shallow, and their bottom seems to be only the continuation, by gentle slopes, of the relief of the continents that border them. Thus the Baltic Sea has a depth of only 120 feet between the coasts of Germany and those of Sweden; scarcely a twentieth part of that of Lago Maggiore, in the Italian Alps; further north, it becomes deeper. The Adriatic, between Venice and Trieste, has a depth of only 130 feet. In these two cases, we see that the bed is only the continuation of the gentle inclination of the plains of Northern Germany and of Friuli. It is the same with the Northern Sea, and with those which wash the British Islands. Here is found a submarine plateau, which serves as a common basis for the coasts of France and the British Islands; nowhere does it sink lower than 600 feet, and frequently it rises much higher. Between France and England, the greatest depth does not exceed 300 feet; but at the edge of the plateau, south-west of Ireland, for example, the depth suddenly sinks to more than 2,000 feet; we may say that here the basin of the Atlantic really begins.

The seas in the south of Europe are distinguished from the preceding by their much greater depths. The basin of the Mediterranean may be called a basin broken through, and fallen in, resembling on a small scale what the Pacific Ocean is on a large one. All the short and abrupt slopes of the lands surrounding it fall rapidly towards the interior. The western basin, in particular, seems to be very deep; it is isolated from the Atlantic, by a submarine ridge or neck, which, in the narrowest part of the Strait of Gibraltar, is not more than 1,000 feet below the surface. But a little further towards the east, the depth falls suddenly to 3,000 feet; and at the south of the coast of Spain, and of the Sierra Nevada, a depth of nearly 6,000 has been ascertained by Captain Smith. Captain Berard indicates still greater depths on the coast of Algeria. If we may believe Marsigli, and if he has not made some mistake in the statement, there has been found in the prolongation of the Pyrenees, the enormous depth of 9,000 feet. Not far from Cape Asinara, on the north-west of Sardinia, the plummet has been sunk, without touching bottom, at a depth of nearly 5,000 feet.

Between Sicily and the coast of Africa, at Cape Bon, a second neck, from 50 to 500 feet in depth, separates the western from the eastern basin of the Mediterranean. The latter seems to be less deep than the former; nevertheless, depths of from 2,000 to 3,000 feet have been determined in the neighborhood of the Ionian Islands, and the southern coast of Asia Minor.

The Black Sea seems to partake of the character of a sunk basin. The Russian maps give it more than 3,000 feet south of the Crimea, and 2,500 on the coast of Abkhasie. The Caspian Sea, placed on the limits of the northern plains, and of the table land of Persia, is composed of two basins. The northern part, as fa as the Caucasus, is shallow; it is the continuation of the low plains of the Volga, and of the Oural. This limit passed, the depth rapidly increases towards the basis of the high chain of the Demavend.

Thus, in the European seas, the depth increases with the elevation of the surrounding lands.

The line of the islands and the peninsulas forming along the eastern coast of Asia the numerous locked seas we have already named, seems to indicate the ancient border of a continent. Within this line these seas have only an inconsiderable depth. The seas which bathe the archipelago of the Sunda Islands, and of Southern China, scarcely anywhere reach the depth of 300 feet. Further north, we find scarcely four or five hundred feet, even at a distance of more than 100 miles from the coasts. The deeps of the ocean begin only outside of the line of the islands.

Since Dampier, it has often been said that the sea is always deep at the foot of high and steep shores, and shallow at the edge of low coasts. The facts just cited prove that this observation, correct in many cases, has only a relative value, and does not hold good universally. Those shallow seas of Eastern Asia are edged in great part by very high lands. The massive point of the south of Africa ends with abrupt coasts, and yet it is necessary to go out more than 100 miles before finding 600 feet of water. According to this rule, we should expect to find no greater depth of sea than at the western foot of the lofty Andes, the declivities of which sweep down so suddenly into the

Pacific Ocean; and nevertheless, under the parallel of Lima, this ocean has only 600 feet, more than forty miles from the coast. On the other hand, the low plains of the Landes of Bordeaux, on the coast of France, lying along the Gulf of Biscay, look out upon a sea, the bottom of which, at a short distance, sinks already lower than a thousand feet.

In Central America, the Gulf of Mexico, 300 miles from the coast of the United States, and 100 miles north of Yucatan, has a depth of only 600 feet; it is, perhaps, a submarine continuation of the plains of Mississippi. Beyond the line of the Antilles, on the contrary, in the volcanic basin of the Caribbean Sea, Captain Sabine indicates a temperature taken at 6,000 feet below the surface.

With regard to the depths of the open sea, they are still but little known; we have, however, some very interesting measurements in the Atlantic Ocean. I cite the most remarkable:

Captains Scoresby and Parry found the basin of the Polar seas very deep, but unequal. Scoresby did not touch bottom, at the 76th degree of north latitude, with a sounding line of 7,200 feet in length. Captain J. Ross went beyond 6,000, in Baffin's Bay. Science is indebted to the skilful direction of the United States Coast Survey, for a great number of very instructive soundings on the American side of the middle part of the Atlantic. One of the ablest of those engaged in this service, Captain Charles H. Davis, U. S. N., whose labors have contributed so much to a better knowledge of the true conformation of the submarine portion of the United States coasts dropped the lead 7,800 feet, about 250 miles south of

Nantucket; Lieutenant G. Bache, 13,000 feet, 34° N. L. The report of Professor A. D. Bache, Superintendent of the Survey, mentions, among the operations of 1848, a thermometrical sounding, taken by the late Lieutenant R. Bache, off Cape Hatteras, giving a depth of 3,300 fathoms, or 19,800 feet, without reaching the bottom. This depth, which left far behind all hitherto ascertained in these quarters, has very recently been surpassed by a sounding executed by Lieutenant Walsh, U. S. N., under the direction of Lieutenant M. F. Maury,* director of the Observatory at Washington. On the 15th of November, 1849, east of the Bermudas, 31° 59′ north latitude, and 58° 43′ 25″ west longitude from Greenwich, in the immediate neighborhood of the position assigned to the rocks called the False Bermudas, the weather being calm and beautiful, Lieutenant Walsh, U. S. N., sunk the lead to the depth of 5,700 fathoms, or 34,200 feet, without touching bottom. The breaking of the line alone prevented him from reaching a greater depth still, for all the circumstances seemed eminently favorable. This depth, which exceeds by 6,600 feet the deepest of the celebrated measurements of Captain J. C. Ross, is the greatest ever ascertained, and reveals, beneath the tranquil surface of the ocean, abysms which we hardly ventured to suspect. The southern basin of the Atlantic seems to have its share of these immense depths, although some indications have given rise to the belief

* For the opportunity of making known this remarkable fact, I am indebted to the kindness of this scientific gentleman, whose zeal for the study of the phenomena of the sea, and the services he has rendered to navigation, by the study of the winds and currents, are universally known.

in the existence of high bottoms, separating this basin from that of the north. Captain J. C. Ross found 16,000 feet in depth, west of Cape of Good Hope, and 27,600, without touching bottom, west of St. Helena. The first of these measurements equals the height of Mt. Blanc. and the second almost reaches that of the Dhavala-Giri; but it would be necessary to add Mt. Washington, with its 6,000 feet, to the height of this giant of terrestrial summits, to attain a height equal to the sounding of Lieut. Walsh. Thus the greatest known sea depth, added to the elevation of the highest mountain of the globe, gives us over 62,000 feet for the thickness of the layer of our globe, upon which our investigations have supplied some information.

Dr. Young, relying upon deductions drawn chiefly from the theory of the tides, thought himself justified in assigning about 15,000 feet to the Atlantic, and about 20,000 feet to the Pacific. D'Aubuisson believes them not to exceed from 9,000 to 12,000 feet. Now we see that the actual measurements leave these estimates far behind.

Laplace, guided by theoretical considerations with regard to the general form of the globe, admitted that the mean depth of the seas is a quantity of the same order with the mean elevation of the continents; which, he says, does not surpass a thousand metres, or 3,000 feet. But the beautiful researches of Humboldt have proved that this estimate of the mean relief of the continents is far too high. He has sought to determine its true value, and has found, as the most probable result, the following numbers for the different continents:

	Mean elevation.
Europe, ...	671 feet.
Asia, ...	1,151 "
North America, ...	748 "
South America, ...	1,132 "

He consequently places the mean elevation of the continental lands at 1,008 feet above the level of the ocean.

Now this number is evidently too low to express the mean depth of the oceans.

It has seemed to me that it would not be uninteresting to compare the depths observed in the Southern Atlantic, with what would be found by supposing that the general planes of inclination of the opposite continents of Africa and America were prolonged until they met under the surface of the ocean. Professor B. Peirce has had the kindness to make the calculation, which gives the following result. Taking as points of departure for Africa, the table lands of Southern Africa, at the foot of the Lupata, and estimating their height at 5,000 feet; and for America, the table lands of Bolivia, estimated at 12,000; the planes which pass through the respective coasts of these continents intersect each other at a distance of nearly a thousand miles from the coast of America; that is, a third of the way across the Atlantic, and 7,600 feet below the surface of the oceans. If the points of departure were taken at the summit of the Lupata, at [illegible],000 feet, and of the Andes, at 24,000, the depth would be about 15,000 feet, which perhaps is not far from the mean depth of this part of the great oceanic valley. But we see that other causes depress the level, in some parts, to depths twice as great. The basins are not, therefore, mere continuations of the general relief of the continents, and this is otherwise shown by their conformation. On leaving the greater part of the shores, the submarine ground descends slowly, in a proportion sufficiently analogous to the general slopes presented by the ground above water on the continents. But, at a point more or less distant from the shore, the slopes abruptly change, the depths suddenly increase, and often become ten times as great at a short distance. I will cite, as examples, the very exact lines of soundings traced perpendicularly to the coast, at several points, between New Jersey and Block Island, at the eastern extremity of Long Island, under the direction of Professor A. D. Bache, for which I am indebted to his kindness. In all the sections, we see the ground descend slowly, gradually, and without great variations, to the distance of 80 or 100 miles from the shore, where the depth scarcely reaches from four to five hundred feet. Beyond this distance, the depth invariably increases so rapidly, that it is sufficient to proceed ten miles further to find three or four thousand feet. The first part is about five feet a mile; the second, more than four hundred.

It is the same with nearly all the grand banks, or high bottoms under water. That of Newfoundland, that of Las Lagullas, at the southern point of Africa, all terminate, towards the depths of the ocean, in abrupt descents, along a great part of their extent. The bottom of the sea, near the coast, and the great banks, present themselves as high plateaus, compared with the bottom of the oceans. But differences of level, amounting to 10,000 feet over a horizontal space of ten miles; of 20,000 feet, as between the sea of Cape Hatteras and the shore; of

34,000 feet, as between the sea of Bermudas and the neighboring continent, reveal to us forms of relief of such a magnitude that we seek in vain for their parallel in the most elevated terrestrial masses. Neither the broad table lands of Mexico, or of California, nor those of Bolivia, or of the Himalaya, which after all are only local swellings, offer proportions approaching those we have just stated in the submarine relief of the basin of the oceans. I must here stop from pursuing this interesting question of comparative physical geography, which at present I barely touch upon. We must wait for still more numerous facts, in order to attempt its complete solution. Let me add, that it would be worthy of an enlightened government, like that of the United States, to bestow on science a regular section across the middle regions of the northern Atlantic basin, — that grand highway of the nations, travelled by thousands of ships. Three or four hundred soundings, penetrating to the bottom of its abysses, would be enough to give us a tolerable idea of the forms of this basin, and would, doubtless, correct many false ideas imposed upon us by the limited views of geology we have gained upon our continents. Many of the scientific expeditions of modern times have had an aim, and have accomplished results less useful to science, and have cost sums vastly greater, than would be necessary to furnish science with such important information.

What the measurements above indicated have settled, may be thus summed up: —

The seas in the neighborhood of the continents are ordinarily of but little depth, and seem to indicate a continuation of the relief of the continents. But at a certain distance from the shores, the sounding gives suddenly great depths, and this abrupt transition seems to indicate the submarine border of the proper basin of the oceans.

Certain interior seas, like the Mediterranean and Caribbean, are deeper than would be expected from their proximity to the lands, and seem to be sunken basins, the form of which is connected with the volcanic phenomena often displayed over their whole extent, but chiefly on their margins; that is, on the principal line of fractures.

The interior of the basin of the oceans is unequal, generally deeper than towards the borders. The greatest observed depths are found in the middle region of the Atlantic. They equal, or surpass, by several thousand feet, the elevation of the highest mountains of the globe, and are found, like them, in the neighborhood of both tropics.

The mean depth of the basin of the ocean seems to be much more considerable than the mean elevation of the continents above their surface, and appears to increase with the relief of the neighboring continents.

We have thus finished our survey of the great forms of the terrestrial surface. We have examined them as the anatomist would examine the body of an animal. This was the first step to take, the necessary condition of our study. But this knowledge cannot be sufficient. We must now see these great organs in operation; we must see them in life, acting and reacting upon each other; we must commence *the physiology of the continental forms.*

But we will not plunge into the infinite details that might be embraced in our subject. We shall continue to study the great features, the prominent traits, which will offer themselves to our notice.

After having recognized great terrestrial individuals, presenting, by their forms and their disposition, an assemblage of characters peculiar to them alone, we have to inquire if, in virtue of these forms themselves, and of this particular position, each of these individuals has not a peculiar physical life, manifesting itself, in the main, by a climate, a vegetation, an animal world, and, relatively to human societies, by special functions which belong to no other. We shall endeavor to discover if there is not here, also, a general law, which gives us the key to all these partial phenomena, helping to group them, and to grasp, in the true point of view, the collective manifestations of the life of our planet, whether in nature, or in the history of man.

But to this end, gentlemen, in order to place you in the point of view from which I would have you consider with me the phenomena of the life of the globe, I cannot avoid the necessity of carrying you for a moment into a world somewhat different from the world of forms wherein we have thus far moved, and to appeal to the eyes of your *mind*, rather than to those which have, up to the present moment, been fixed upon these maps.

In fact, nothing less is necessary than to say to you, in as few words as possible, — to prove to you, if it can be done, — that there is a law of life and of growth, which, if taken in its most general formula, in its *rhythm*, is applicable to all that undergoes the process of development.

All life, as we have said, gentlemen, in its most simple formula, may be defined as a *mutual exchange of relations.*

An exchange supposes at least two elements, two bodies, two individuals, a *duality* and a difference, an inequality between them, in virtue of which the exchange is established.

There is, then, at the foundation of all the phenomena of life, a difference between two or more individuals, calling out an action and reaction of one upon the other, the incessant alternation of which constitutes the movement we call life, and which gives birth to all the phenomena we consider as its manifestation.

Let us endeavor, first, to detect this law in inorganic nature.

This lamp that gives us light, the gas that burns before our eyes, what else is it than one of those phenomena of inorganic life, the result of the mutual and repeated action of two heterogeneous bodies upon each other? We have, on the one side, the hydrogen gas, conducted by this pipe, and brought into the presence of oxygen contained in the air. These are two bodies considered as simple, but having different properties. Place them in contact, under suitable conditions of temperature, and the mutual action immediately commences; they combine with an activity which becomes visible to the senses by the rapid development of heat and light; and in this continuous, vital movement.

their differences are extinguished, or rather combine and harmonize in a new body, a product, the end of all this activity, in which the antagonism of the primitive elements has ceased. This new body is water; it is a liquid, and no longer a gas; it is a body, all the physical properties of which are different from those that compose it, which, as you know, play very different parts throughout nature. This same gas that serves to light us, contains also carbon; this also combines with oxygen to form a new body of carbonic acid gas, the properties of which are all special in it.

Each of these new products may, in turn, enter into relations of exchange with others, and pass as an elementary body into a new combination, the result of which will be a body composed of four simple elements, but endowed, as such, with entirely different qualities, belonging to it alone. It may, in turn, become one of the elements composing a multitude of bodies; and it is thus that the sixty elements our chemical means have not enabled us to decompose, which chemists call simple bodies, supply nature with materials sufficient for the immeasurable variety of all the compound bodies that exist.

What do we see, finally, in all this physical and chemical process? A primitive difference between two substances, an action and reaction of one upon the other, and their combination in a new body, which may, in its turn, perform the same part. I mark, gentlemen, these phases of the phenomenon going on, at the present moment, under our eyes.

Without coming into combination, a difference between

two bodies excites none the less a vital movement. Place near each other a plate of zinc and a plate of copper; these two enter immediately into an interchange of positive and negative electricity, and give birth to those powerful electrical and magnetic currents which modern industry puts to such admirable use. I say, further, place side by side two plates of the same metal, but unequally heated, and there is established between them an interchange of temperature, and of electrical currents of the same nature. Thus everywhere a simple difference, be it of matter, be it of condition, be it of position, excites a manifestation of vital forces, a mutual exchange between the bodies, each giving to the other what the other does not possess. To multiply these differences, to increase their variety, is to render the actions and reactions more frequent, is to extend and to intensify life.

But let us pass to organized nature. It would be easy to demonstrate, gentlemen, that the law we have just recognized is also that which governs the growth of the vegetable; but I would rather trace it in the animal world, wherein it is expressed still more clearly.

Let us see, first, how nature proceeds in the formation of the organic individual, the animal. No one has shown it better than my learned friend,* whom I need not name in this place. Thanks to him, these facts have become familiar to you; I shall need only to recall them to your minds.

I begin with the animal considered in itself as an

* Professor Agassiz.

individual. In a liquid animal matter, without precise form, homogeneous, at least in appearance, a mass is outlined which takes determinate contours, and is distinguished from the rest; it is the egg. Soon, in the interior of the egg, the elements separate, diverging tendencies are established; the matter accumulates and concentrates itself upon certain points; these accumulations assume more distinct forms and more specific characters; we see organs traced, a head, an eye, a heart, an alimentary canal. But this diversification does not go on indefinitely. Under the influence of a special force, all these diverse tendencies are drawn together towards a single end; these distinct organs are united and coördinated in one whole, and perform their functions in the interest and for the service of the individual commanding them.

What, then, has been the course pursued here by nature?

The point of departure is a unit, but a *homogeneous unit,* without internal differences; a chaotic unit, if I may venture to say so; for what is a chaos but this absence of organization in a mass, all the parts of which are alike?

The progress; it is *diversity*, the establishment of differences, the giving to forms and functions their special characters.

The end; it is a new unit, *the organic or harmonic unit,* if you please; for all the individual organs are not fortuitously assembled, but have each of them their place and their functions marked out.

The totality of these evolutions is what is ordinarily called *development.*

The progress, we say, is *diversification* it is the variety of organs and of functions. What, tnen, is the condition of a greater amount of life, of a richer life, of a completer growth for the animal? Is it not the multiplicity and the variety of the special organs, which are so many different means whereby the individual may place himself in relation with the external world, may receive the most varied impressions from it, and, so to speak, may *taste* it in all its forms, and may act upon it in turn? What an immense distance between the life of the polype, which is only a digestive tube, and that of the superior animals; above all, of man, endowed with so many exquisite senses, for whom the world of nature, as well as the world of ideas, is open on all sides, awakening and drawing forth, in a thousand various ways, all the living forces wherewith God has endowed him!

And what we here say of organic individuals — is it not true of societies of individuals, and particularly of human societies? Is it not evident that the same law of development is applicable to them? Here, again, homogeneousness, uniformity, is the elementary state, the savage state. Diversity, variety of elements, which call for and multiply exchanges; the almost infinite *specialization* of the functions corresponding to the various talents bestowed on every man by Providence, and only called into action and brought to light by the thousand wants of a society as complicated as ours — these have, in all times, been the sign of a social state arrived at a high degree of improvement.

Could we, indeed, conceive thy possibility of this multitude of industrial talents that have their birth in the wants of luxury, and are revealed by the thousand elegant nothings displayed in our drawing-rooms, among the Indians of the Rocky Mountains, sheltered by the few branches which form their wretched huts? The commercial life, which creates the prosperity of the foremost nations of the globe,—is it possible among a people whose ambition is limited to hunting in the neighboring wild the animal that is to furnish food for the day? Could we hope to see the wonders of architecture unfolded among a people who have no public edifices but the overhanging foliage of their forests? Had Raphael been born among them, would he ever have given his admirable masterpieces to the world? And the precious treasures of intelligence and of lofty thoughts contained in our libraries,—where would they be, if human societies had preserved the simplicity a false philosophy has called the simplicity of nature, but in reality the most opposed to the true nature of man?

No, gentlemen, it is the exchange of products by the commerce of the world, that makes the material life and prosperity of the nations. It is the exchange of thoughts, by the pen and by speech, that sets in motion the progress of intelligence. It is the interchange of the sentiments and affections, that makes the moral life and secures the happiness of man.

Thus, gentlemen, all life is mutual, is exchange. In individuals, as well as in societies, that which excites life, that which is the condition of life, is *difference.* The progress of development is diversity; the end is the *harmonious unity* allowing all differences, all individuualities to exist, but coördinating and subjecting them to a superior aim.

Every being, every individual, necessarily forms a part of a greater organism than itself, out of which we cannot conceive its existence, and in which it has a special part to act. By performing these functions, it rises to the highest degree of perfection its own nature is capable of attaining. Unhappy he who isolates himself, and refuses to enter into those relations of intercourse with others which assure to him a superior life. He deprives himself voluntarily of the nutritive sap intended to give him vigor, and, like a branch torn from the vine, dries up and perishes in his egoism.

All is order, all is harmony in the universe, because the whole universe is a thought of God; and it appears as a combination of organisms, each of which is only an integral part of one still more sublime. God alone contains them all, without making a part of any.

LECTURE IV.

Recapitulation—Is the law of development applicable to the whole globe, considered as an individual?—Origin of the Earth according to the hypothesis of Laplace and Herschell—Gradual formation of the continents—Europe at the Silurian epoch—North America at the Carboniferous epoch—Character of inferiority of the organized beings which correspond to these ancient formations—Europe at the Tertiary epoch—Greater diversity and perfection of the organized beings—Distinction of the three epochs; the insular, the maritime, and the continental—The formula of development the same for the entire globe and for the organized beings—Consequences—The law of differences and the law of contrasts—The three grand terrestrial contrasts.

LADIES AND GENTLEMEN:—

We have recognized, in the life of all that develops itself, three successive states, three grand phases, three evolutions, identically repeated in every order of existence; a *chaos*, where all is confounded together; a *development*, where all is separating; a *unity*, where all is binding itself together and organizing. We have observed that here is the law of *phenomenal life*, the *formula* of development, whether in inorganic nature or in organized nature.

The differences are the condition of development; the mutual exchanges, which are the consequence of these differences, waken and manifest life.

The greater the diversity of organs, the more

active and the superior in its nature is the life of the individual.

The greater the variety of individualities and relations in a society of individuals, the greater also is the sum of life, the more universal is the development, the more complete, and of a more elevated order.

But it is necessary, not only that life should unfold itself in all its richness by diversity, but that it exhibit itself in its utility, in its beauty, in its goodness, by harmony.

Thus we recognize the proof of the old proverb, "variety in unity is perfection."

If such is the law of life in all beings, it ought equally to be the law of life of our entire globe, collectively considered, as a single individual. It is the investigation of this question I am going to attempt this evening.

The investigation, in order to be complete, would presuppose a perfect knowledge of the origin of our globe. But who is ignorant that in this respect we are yet in the world of suppositions? Nevertheless, the brilliant hypotheses of Laplace and Herschell on the primitive formation of our planet, and the results, better founded, perhaps, geology gives us, upon the history of the successive changes the surface has undergone, permit us, if I do not deceive myself, to detect with certainty the great phases of development we wish to ascertain. I am aware of the objections that may be made to both the one and the other; but it seems to me that they bear more upon the details, than upon the fundamental facts, and that in astronomy, as in geology, certain great truths are none the less gained for our knowledge. Now it is precisely these general facts that proclaim, in a language perfectly clear, the reality of the law of development we have endeavored to illustrate by the preceding examples.

Laplace, Herschell, and most other modern astronomers agree in considering the assemblage of stars that form, at present, our solar system, as having been at the first confounded in one celestial body, resembling one of those mysterious nebulæ we see floating in the celestial spaces. This nebula would have a solid and luminous nucleus or core, if we ascend no further in this history than the point to which we are led by the hypothesis of Laplace. But if, by the help of the analogies drawn from the celestial bodies collectively, we scale still higher, with Herschell, towards the probable origin of the world, we shall be able to conceive it as being entirely gaseous, and even as forming a part of the general matter, spread uniformly throughout space. A gaseous mass, uniform, or rather formless—for the property of gas is indefinite expansion—an obscure mass, where nothing is determined, this is chaos, this is the inorganic state, here is the point of departure.

But soon the development begins. A principle of concentration,—gravitation,—counterbalances the unlimited expansion of the gaseous matter, brings the molecules nearer together, and groups them in a spheroidal mass. This approximation allows the molecules, different in nature, to act upon each other according to their chemical affinities; the process of life commences, and its earliest manifestation is light and heat. The nebula is detached from the general mass under the form of a luminous spheroid, traced in the obscurity of the heavens. This is the first step in the process of formation.

This gaseous spheroid then resolves itself into local agglomerations, which, while concentrating each in itself, under the influence of gravitation and chemical combinations, separate from each other in distinct spheres. Whether this phenomenon is effected, as Laplace imagines, by the successive separation and agglomeration of concentric layers of the solar atmosphere, or in virtue of some organic law, still unknown, is of little importance here. The fact of the separation of the different bodies of our solar system into a number of spheres, planets, and satellites, is not less certain, and constitutes one of the essential and incontestable phases of its development.

Let us leave the other stars, elder and younger brothers of the earth, and follow henceforth the ulterior changes our own globe undergoes.

The gradual concentration, and perhaps certain changes of temperature, permit successively the combination of a multitude of different bodies; and the result, as far as regards the general forms, of all this mighty chemical and physical life, is to present matter, no more under a single form, gas, but under the three forms of gaseous, liquid, and solid matter. These three elements ranging themselves in the order of their density, the globe is composed of a solid mass at the centre, enveloped first by a liquid, and secondly by a gaseous covering—the ocean, and the primeval atmosphere.

At the surface, meantime—and it is the history of this surface which it concerns us most to know for our study—two elements only are in contact, the air and water. The winds and the marine currents, owing to the unequal distribution of the solar heat, doubtless exist; but the differences of temperature being very inconsiderable between one place and another, they must be languid, and, besides, they are perfectly regular and uniform; for no land disturbs the equilibrium of the temperature of the atmosphere due to this general cause, or interrupts or breaks the course of the currents. On account of a density of the atmosphere probably greater, and perhaps of a higher degree of heat in the globe itself, the temperature is more uniform from one end of the globe to the other. The rains, if the state of the atmosphere permits their existence, are useless, for there is no land to receive them, and to render them subservient to life. In this state of things, organic life is nevertheless possible. Plants and animals live in the bosom of the ocean; but the earliest fossiliferous strata, which doubtless represent this epoch, contain none of either, except a few types but little varied, and all belonging to the lowest grade in the scale of organized beings. It is the dawn of life, the infancy of the vegetable and animal kingdom.

A new difference is now added, and marks a new progress. In the train of internal movements, or rather, by the effect of a simple cooling of the globe, the third element, the *solid*, the earth, quitting the place its weight had assigned to it, rises from the bed of this boundless ocean; it lifts itself above the level of the waters, cuts the surface, puts itself in contact with the atmosphere, from which it had been separated by the whole thickness of the primeval ocean, and warms itself in the life-giving rays of the sun.

This fact of the appearance of the firm earth above the waters of the oceans, is an immense step in the rise and growth of the life of the globe. The three forms of matter react henceforth upon each other; the atmosphere, the seas, and the lands, absorbing the solar heat in an unequal manner, the ancient equilibrium is destroyed; the winds, the currents, are modified in their march; the climates are more varied; the rains become useful, and henceforth water and fertilize the land. Finally, a new element renders the appearance of a greatly superior organic life possible, and becomes the seat of a vegetation, and an animal world of a very different degree of perfection from that which existed before. It is a victory gained by higher life over matter, which it compels to serve a more exalted end.

But geology demonstrates that in the earliest ages of the epoch of organic life on the earth, the organic epoch, as I would fain call it, the firm lands are reduced to a few islands only, scattered over the bosom of the oceans.

"Apparent rari nantes in gurgite vasto."

Everywhere the beginnings are modest. The place of the future continents is not yet marked, except by a few scattered stripes, forming here and there a few archipelagos. It is the *insular* epoch comprising all the earliest ages of geology You will see this by the two maps before you; the one represents Europe at the Silurian epoch, the most ancient of the fossiliferous strata, and the other, North America at the Coal epoch, which, although a little more recent, belongs almost to the same age. (See Figs. 2 and 3.)

Fig. 2.

Europe at the Silurian Epoch.

It is doubtless hardly necessary to state that such maps can only be approximations. They indicate substantially those of the present dry lands which already existed at that time, and which have not been covered by the waters of the ocean since those ancient epochs, except, perhaps, in the diluvian. But imperfect as are the data of geology, in this regard, the fact of the gradual increase of the dry lands is none the less placed beyond a doubt.

The largest domain, then, above the surface of the water, in the regions of the future continent of Europe, was Scandinavia, and a part of Russia. England and Scotland are only marked by a few islands along the existing western coast; Ireland, by a few others, placed at the corners of the present island. All France is represented merely by an island, corresponding to the central table land of Auvergne, and by some strips of land in Vendée, in Brittany, and in Calvados. In Germany, Bohemia forming a great island, the Harz, and the plateau of the Lower Rhine; small portions of the Vosges, and of the Black Forest, and some low lands on the spot occupied by the Alps, between Toulon, Milan and Tyrol, compose an archipelago which is to become the centre of the continent. All the regions of the South, except, perhaps, a few small portions of Spain and of Turkey, do not yet exist.

North America, at the epoch when the coal deposits are formed, is, in like manner, made up of a few islands only, analogous to Scandinavia, but less numerous, less parcelled out than we find them in Europe at the same period. A large island occupies all the present north-east of the continent, with the region of the Alleghanies and the Apalachian, and all the region north-west of the

Valley of the Mississippi, and forms a species of small continent, in the interior of which are three large inland seas, or three large swamps, where the plants are vegetating that compose the great coal deposits of the present day. A similar sea doubtless lay between Nova Scotia and Newfoundland, bordered, perhaps, by lands which have disappeared beneath the waves. All the great belt of low lands along the Atlantic coast and the Gulf of Mexico, including Florida, did not exist; the ocean formed a deep gulf, running up the Valley of the Mississippi one half its length.

Fig. 3.

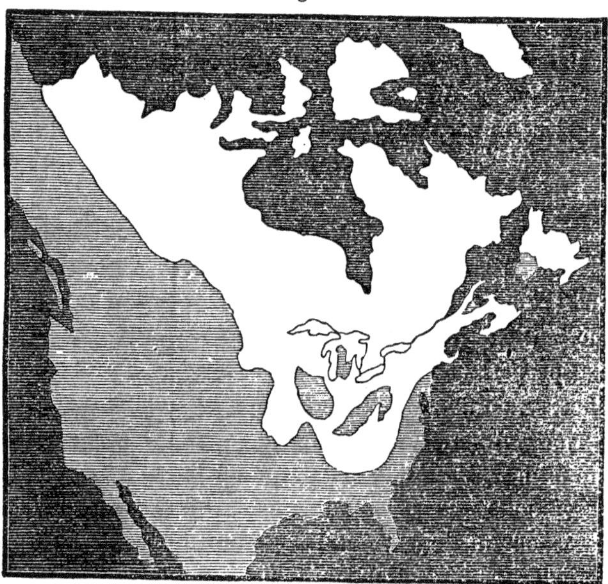

America at the Coal Epoch.

The vast plains west of the Mississippi, the Rocky

Mountains, the table lands and the high snow-capped chains from California to the Frozen Ocean, were still at the bottom of the sea.

This augmentation of the number of the islands, their clustering in archipelagos, is certainly a progress; there is still, however, but little variety; the mountains are few in number, and slightly elevated; the valleys traced but indistinctly or not at all; the slopes imperfectly determined; extensive low and swampy regions indicate still the preponderance of the watery element. A thicker and denser atmosphere equalizes the temperatures. One species of climate alone, the maritime or insular climate, moist, without extremes, reigns over land and sea. No great continents, none of those elevated masses which give to climate extreme and variable temperatures, and the character of dryness; none of all those varied forms of vegetation which show themselves later under its influences.

The organized beings corresponding to this physical condition of the surface of the globe, show with the utmost clearness this character of uniformity and inferiority. From one extremity of the earth to the other, the Trilobites of the Silurian epoch are found identical in their species, at once in America, in Europe, in Africa, and in New Holland. The vegetables, accumulated in the coal beds, are the same at the poles and the equator. The types of organized beings are not only few in number, but they all still belong to those which mark the inferior degrees of animal life; and in each class, from the radiates to the fishes, the highest beings of this primitive creation, the prevailing forms are those that characterize the embryo in the early periods of its development. Such are the numerous Crinoids, the Brachiopods, the Trilobites; and, in the fishes, the Ganöids of the Silurian epoch. Such are still the great Entomostraca of the carboniferous epoch, and in the vegetables the gigantic ferns, the horse-tails, (Equisetaceæ,) the palm trees, and the coniferous trees, the accumulated remains of which compose the immense beds of coal, provident nature has deposited for the present and future wants of human industry. The two first of these vegetable types belong to the inferior order of the cryptogamous plants; the third to that of the monocotyledons; the fourth, the coniferous, is scarcely placed higher.

We must abstain from pursuing here in its details the admirable history of the surface of our earth, and of the new beings which successively appear; this is the business of geology. Let us say, only, that one of the most beautiful of these results is the demonstration that the diversity of terrestrial forms, the variety of the types and species of organized beings, become always greater and greater. Every new revolution is a new progress; we see one elevation added to another; one surface after another emerging to increase the existing dry lands; one chain of mountains after another appearing and binding together the hitherto separate islands. The terrestrial masses enlarge in number and size; their contours are more varied, their surfaces more broken up.

Let us cast our eyes upon this chart, representing Europe at the commencement of the Tertiary epoch. Comparing it with the map of the Silurian epoch, we shall be able to form an idea of the change that has been wrought. (See Fig. 4.)

Fig. 4.

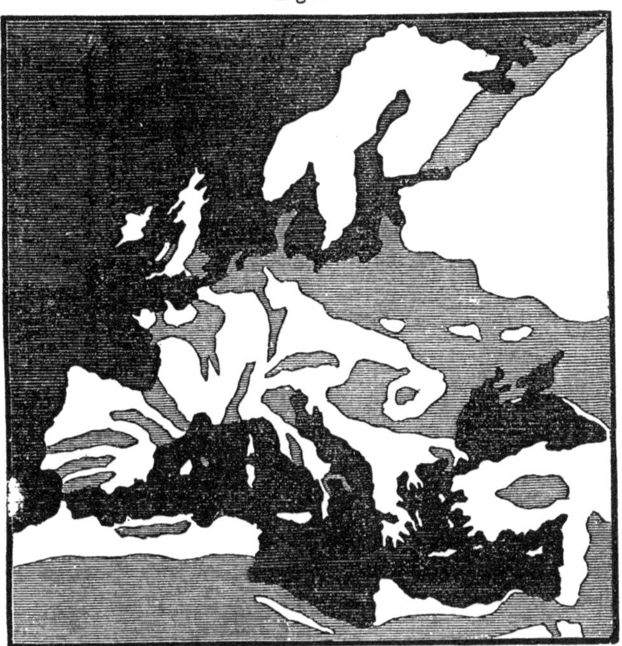

Europe at the Tertiary Epoch.

Not only the number of the lands has been multiplied, but everywhere the primitive islands have been enlarged and consolidated. The centre of the continent, Germany and France, constitutes already a considerable collective region, unbroken save by a few interior basins. The British Isles form already two or three large islands, and the eastern part of England only is wanting. The three peninsulas of the south are clearly traced; Italy only is still exposed along its coast to the encroachments of the sea; Scandinavia continues to form a large solitary island; the mountains are more elevated; the Pyrenees, the Appenines, a small part of the Alps, already mark out the great features of relief which characterize the continent.

During the tertiary epoch, the variety of physical circumstances is still increasing; a multitude of isolated basins, like those of Paris, of London, of Oeningen, assume a special physiognomy, and have their separate faunas. The natural physical regions are determined, and take their distinctive character. The climates are diversified with all the physical circumstances of a country, and are reflected in the ever-increasing diversity of animal and vegetable genera and species.

Meantime, this movement of *specialization* is not going to extremes. The masses of earth, while becoming more numerous, more various, more diversified in shape, are grouping themselves more and more; the contours of the continents are getting better defined; the tertiary basins are filling and drying up. The water of the seas disappearing from the interior, the atmospheric waters, which run on the surface, supply their place, scoop out their valleys, make the slopes regular, equalize the soil by spreading over it their precious alluvium. The diluvial torrents and the immense glaciers, contemporaneous with this epoch, complete the shaping of the soil and the preparing of this fertile loam, which will richly repay the toil of the laborer. The earth is ready to receive its lord.

It is thus, by a process of admirable simplicity, this

diversity of successive elevations is combined into a few great units, a few continents; these in turn are grouped in two worlds and form an organism, with some of the features of which we have already become acquainted.

This same progress is confirmed by palæontology, through all the successive ages of nature. The variety and the perfection of the types and species keep pace with the increasing diversity of the lands and the seas, and all the physical circumstances which serve as the basis and the condition for the life of plant and animal. In the *insular* or oceanic epoch, that of the palæozoic strata, we have seen animals entirely marine prevailing, and forming the inferior and embryonic types of the four divisions of the animal kingdom; it is the reign of the fishes, if we take the vertebrates as the type of development. During the formation of the secondary strata, which I would call the *maritime* epoch, on account of the great land-locked seas that characterize it, the huge reptiles, the monstrous Saurians of the Jurassic waters, are the prevailing form, and by their amphibious habits mark at once their more elevated position in the animal scale, and the increasing force of the land element. The numbers of living genera and species are much greater than at the palæozoic epoch, but the same types are still uniformly spread over vast spaces.

The tertiary epoch, which I would call the *continental* epoch, beholds the appearance of the superior animals, the mammifers, the life of which is almost exclusively attached to the solid land. The continental element triumphs; all the faunas become localized; each country of the globe has its appropriate animals; the variety of animal and vegetable species grows almost to infinity. But the unity reappears with the creation of man, who combines in his physical nature all the perfections of the animal, and who is the end of all this long progression of organized beings.

If we cast a glance back upon the way we have just passed over, do we not, gentlemen, recognize a striking analogy between this successive formation, first of our solar system, then of the continents and the beings inhabiting them, and the formation of the animal in the egg? Is there not here the same law that we have recognized everywhere else? Do we not see, first, a homogeneous fluid, then the appearance of elementary organs at several points; finally, their definitive combination in an organic whole? Yes, gentlemen, there is between the two series of facts all the difference of organic and inorganic nature; but the *formula* of development is the same.

The consequences of this fact are numerous; let us point out the most important, those which are chiefly useful for our subject.

1. The law of development is applicable to the land, and to the continental forms.

2. In this order of facts, as elsewhere, the condition of a more active life is a greater variety of forms of nature, of relative situations; in a word, of more varied contrasts.

3. Then other things being equal, we may consider, in advance, those continents as the best endowed, the best organized, the best prepared for the development of human societies, which present the most varied contours,

the most diversified forms, the most numerous contrasts, and the best characterized natural regions. There is here the same relation as between the inferior animal without special organs, and the superior animal richly furnished with special organs.

4. The result of all these differences of forms, of climate, of vegetation, of all the internal and external contrasts, considered in each of the great terrestrial masses, in each continent, is to impress upon every one a special character, a peculiar life, so that they appear as so many individuals, differing from each other, and designed to enter into relations of intercourse and of reciprocal influence.

5. Considered under various aspects, in the point of view of their analogies and their differences, the great terrestrial masses are combined in groups of continents, according to characteristics of the same nature. Now these groups, compared together, present an assemblage of distinct and opposite characters, and seem to form *great contrasts*, two by two. Thus the two continents of America have, notwithstanding the immense differences between them, certain common characters, binding them into a natural group, distinguished, as such, from the Old World, with its three or four continents. It is the same with the three continents of the north, compared with the three continents of the south.

Terrestrial life, if I may say so, is then developed under the influence of a law which we might name *the law of differences;* and in the general phenomena of the life of the globe all partial differences being combined into grand differences, opposed two to two, we may call it *the law of contrasts.*

It is, then, under the form of great contrasts, the sources of a multitude of vital actions, that we shall henceforth consider the continental masses. Now, let us point out the three most important of them.

1. The contrast of the continental hemisphere, and the oceanic hemisphere, or land and water.

2. Of the Old World and of the New World.

3. Of the three northern continents and the three southern continents.

In studying the globe in this point of view, we shall see it under a new light. I know I am stepping a little out of the beaten paths. This is not, believe me, gentlemen, the result of a passing momentary glance, but of patient studies in detail, in the realm of nature and of history. We shall have the pledge that it is not without value, when we trace in each of these contrasts all the great analogies and differences we have thus far shown, each in its place, in its true light, and with the just portion of influence they are respectively entitled to have.

But in setting forth these contrasts, this antagonism of one half the globe against the other, let us hasten to say that there is nothing hostile in the conflict; for it tends to life, not to death. True victory is not to crush an opponent, but to make him a friend. We suspect then in advance, — the law of life declares it, — we suspect, in advance, that all these oppositions resolve themselves into a grand harmony, wherein each continent has its part to perform, while all live at the same time a common life. But to arrive at this final result, nature alone is not sufficient; there is needed something more than a physical tie between all these parts of the world; there is needed a moral bond; a soul is wanting to this body, to set its organs in action. Now, it is man, it is human societies, that alone can animate the great frame, bind together all the parts, and render perfect that organism which is the end and aim of the long procession of existence upon this earth.

LECTURE V.

The north-east or continental hemisphere, and the south-west or oceanic hemisphere — Land and water — Differences in the forms of their surfaces — Continental climate and sea climate — Their different influences upon the vegetation and organized beings — The oceanic the inferior element: the terrestrial element the superior — Blending of the two natures — Transportation of the waters of the ocean into the continents — The atmosphere the mediator between them.

Ladies and Gentlemen: —

We have explained the reasons that have led us to conceive the general phenomena of the life of the globe as taking their source in certain great contrasts, of which we have especially distinguished three: the continental hemisphere opposed to the oceanic; the Old World to the New; the three northern continents to the three southern.

We shall this evening commence the investigation of the first, comparing the terrestrial element with the oceanic element, in order to ascertain the special character of both. We shall then inquire by what means, and to what extent, they enter into relations combining and modifying reciprocally their nature. We shall see, finally, but only by-and-by, the happy and important effects of the contact and blending of the land and the water.

In speaking of the distribution of the lands on the surface of the globe, we have already said that the most characteristic and general trait in this respect is their prevailing concentration in one hemisphere alone, so that the whole world is divided into two great regions; one over which the ocean reigns, the other which the terrestrial masses command by their number, their size, and their connection. Now, the equator divides the globe into a northern and a southern hemisphere, very different, in this respect, from each other. But if, with Ritter, we draw a great circle, passing at once over the western coast of Peru and the peninsula of Malacca, at the southern extremity of Asia, the contrast is more complete still, (See Fig. 1,) and the globe is cut into a north-eastern hemisphere, (if we take the Old World for the eastern,) comprising five of the continents, and the largest; and a south-western hemisphere, where we see hardly anything but oceans, and, in the midst, floating solitary, the most insular of the continents, Australia. The southern point of America and the islands of the Pacific Ocean are the only lands which, with this continent, represent the continental element in this oceanic world. Thus, one of the sides of our planet is the humid, aqueous side; the other is the terrestrial side.

Let us trace, meantime, the characters which distinguish these two elements, in the point of view of physical geography, beginning with the exterior forms.

By its very nature, the liquid element has no form peculiar to itself, except the spherical form of the drop of water. Upon a globe like the earth, the waters of the ocean appear to our eye as a plain uniform surface, which we are even accustomed to consider a constant level, to which we may refer, as to a fixed plane, all the elevations of the firm earth. Their movable molecules undergo all the forms that are impressed upon them by the solid forms with which they come in contact, not preserving any shape after the contact has ceased. All the indentations of the shores of the oceans belong, as we have said, to the continents which they bathe with their waters. On the surface, none of those diversities of relief, enlivening and varying, to infinity, the aspects and the physiognomy of the continents; none of those high mountain ridges, from the summit of which the eye embraces at a glance a portion of our globe as vast as the range of vision. We speak of the sentiment of the infinite which the ocean awakens in the soul of the voyager; but the infinite, it must be confessed, exists only in our imagination, for it is limited to a circular surface of twenty or thirty miles radius. In spite of the proverbial inconstancy and mobility of this element, in spite of the varied movements of its billows, originating in the conflict with the atmosphere, we must acknowledge, gentlemen, that the strongest and most universal sentiment which, on the whole, it inspires in the man who trusts himself to the waves, is that of a despairing monotony. I wish for no other proof of this than the feverish impatience that seizes upon even the mariner himself in the midst of a long calm, and the thrill of joy the first cry of land, raised by the sailor on the watch at the mast-head, excites in the hearts of all. Thus the life of the seaman, so poetically sung, has it not one aspect alone, which is nearly the same, over and over again, from one end of the world to the other?

A difference no less important, between the seas and the lands, is that of the climate. It is owing substantially to the peculiar physical properties of the water and the soil of the continents. Water has a great capacity for heat, but a feeble conducting power; it grows warm but slowly in the rays of the sun. The evaporation being considerable, produces a cooling, which tempers further the heat received at the surface. Finally, the cooler particles of the lower layers, set in motion by the waves and the currents, incessantly fill the place of those of the superficial layer, and prevent it from rising to a high temperature.

It is the same with the cooling. The superficial layer growing cool, whether by the absence of the sun, or by contact with a colder atmosphere, the cooled molecules become more heavy, fall lower, and give way to the warmer molecules of the inferior strata. This motion is incessantly repeated, and singularly retards the process of cooling.

Thus the heating and cooling are less sensible and more slow, and do not reach the extremes. The air itself, by its perpetual contact, shares in the uniformity of temperature which belongs to the surface of the waters, and which, combined with the abundance of vapors that saturate the atmosphere, gives to the sea climate its true character.

It is quite different with the surface of the soil, whose particles are fixed. The soil rapidly absorbs the solar rays; the surface layer is the more heated, since it cannot be displaced, as in the water, by another, and it soon attains an elevated temperature. But, for the same reason, the ground easily loses heat by radiation, whether during the nights or the clear days; and the loss is so much the greater, as the radiation is favored by the inequality of the surface, and the transparency of an atmosphere more dry and less charged with clouds The lands removed from the influence of the oceans have, then, a climate characterized by the extremes of cold and heat, by more violent changes, and a drier atmosphere. These are the essential features of the continental climate. If the former is *constant*, the latter is *excessive.*

If we now observe the manner in which sea and land are affected with regard to their temperature when near each other, and receiving the same quantity of heat from the sun, we notice that the sea is colder than the land during the day, and warmer during the night. In the same way, taking the different seasons of the year, in summer the sea is colder than the land, in winter it is warmer. It preserves the mean temperatures, while the land experiences the extremes. It tends to soften all the differences, to establish uniformity of climate.

A comparison by examples, as far as possible, of the climate of the pelagic islands, subjected to the influence of the surrounding ocean, with the climate of places in the interior of the lands, will bring this difference prominently out. I purposely choose places situated, two by two, in similar latitudes, and successively in latitudes more and more near to the tropics. I ask your permission to cite the numbers in degrees of the centigrade scale, which it would be very desirable to adopt uniformly, at least in matters of science, if not in the common usage of daily life. Nevertheless, for the sake of clearness, I will add the corresponding value in degrees of Fahrenheit. The differences, which are here the important thing, are found in two separate columns.

Let us first compare the climate of the Färoe islands, situated in the midst of the Atlantic, with that of Petersburg, and, if you please, of Yakutsk, in the depths of Siberia; and, to form an idea of the extent of the thermometrical variations each of these climates undergoes, let us establish the difference between the mean temperature of summer and winter, in each of them. These three localities are situated in the high latitudes, between 60° and 62° north lat.

		Winter.	Summer.	Difference deg. Cent.	Difference, deg. Fah.
Färoe, . . .	Cent.	3.6	12.2	8.6	
	Fahr:	38.5	54.0	. . .	15.5
Petersburg,	Cent.	-8.7	16.0	24.7	
	Fahr.	16.3	60.8	. . .	44.5
Yakutsk, .	Cent.	-38.9	17.2	56.1	
	Fahr.	-38.0	63.0	. . .	101.0

We see, by the rapid increase of the differences, how the variations augment, in proportion as we advance into the interior of the continents.

If we compare the mean of the coldest with that of the hottest month in the same places, the proportion becomes still more sensible.

		Coldest Month.	Hottest Month.	Difference deg. Cent.	Difference, deg. Fahr.
Färoe, . .	Cent.	2.7	12.8	10.1	
	Fahr.	36.8	55	. . .	18.2
Petersburg,	Cent.	-10.3	16.9	27.2	
	Fahr.	13.5	62.4	. . .	48.9
Yakutsk, .	Cent.	-40.5	20.3	60.8	
	Fahr.	-40.9	68.5	. . .	108.5

The extremes of temperature differ even more, just as was to be expected. The highest degree of heat observed at Färoe is only 13.5 Cent., or 56.3 Fahr., and it freezes but little there, while the meteorological annals of Petersburg indicate heats of 33.4, and cold of -34.0; that is, extremes 67.4 Cent. or 121 Fahr. apart. It is at once the cold of the poles, and the heat of the tropics. At Yakutsk the mercury remains frozen often for whole weeks, implying a continued cold of at least 40° Fahr. below zero.

Finally, the variations in the same day follow the same relative course; while at Färoe they are scarcely a few degrees, it is not unusal to see, at Petersburg, violent changes of from 30° to 40° Fahr. in the same day.

In the lat. of 50° to 52° N., we find, at Penzance, on the south-west coast of England, and Barnaul, at the foot of the Altai, in Siberia, the following temperatures:

		Winter.	Summer.	Difference, deg. Cent.	Difference, deg. Fahr.
Penzance, .	Cent.	7.0	15.8	8.8	
	Fahr.	44.6	60.4	. . .	15.8
Barnaul, .	Cent.	-14.1	16.6	30.7	
	Fahr.	6.6	61.9	. . .	55.3

The differences, as we see, are still considerable, but less than between Färoe and Yakutsk.

Nearer the tropics, the climate of Madeira, compared with that of Cairo, in lat. 32° and 30° N., indicates a similar proportion.

		Winter.	Summer.	Difference, deg. Cent.	Difference, deg. Fahr.
Madeira, .	Cent.	16.3	21.1	4.8	
	Fahr.	61.3	70.0	. . . .	8.7
Cairo, . . .	Cent.	14.7	29.2	14.5	
	Fahr.	58.5	84.6	. . .	26.1

The differences between the seasons become less in the two localities respectively; but the influence of the ocean and the continent is always very marked. The difference between the extreme temperature, at Madeira only from 12° to 15° C. or 20° to 27° F., is in Egypt 31° C. or 56° F.

In the Sahara, ice has been known to form by the intensity of the radiation, and the heat to rise, by the wind of the desert, to the enormous height of 118° F., or 48° C.

I will only cite one example more, and it shall be taken from the coasts of America, between 31° and 32° N. latitude.

		Winter.	Summer.	Difference, deg. Cent.	Difference, deg. Fahr.
Bermudas,	Cent.	15.1	24.0	8.9	
	Fahr.	59.2	75.2	. . .	16.[illegible]
Natchez, .	Cent.	10.0	25.4	15.4	
	Fahr.	50.0	77.7	. . .	27.7

If the climate of Natchez is less extreme, it is because this place is too near the ocean.

We see, by these tables, how great is the influence of the sea upon the distribution of the temperature in the different seasons of the year, and in the course of the day. It tends to bring the extremes together, and to maintain at all times an equality of temperature.

The sea climate is then equal, uniform, moist; the sky often cloudy and rainy in the high latitudes. The land climate is excessive, unequal, with violent changes, dry; the sky is usually clear.

The astronomical climate, caused by the latitude, is then greatly modified by the presence or absence of the seas; and the distribution of heat through the year, for any place whatever, depends essentially on its proximity to, or distance from, the oceans, and the relative frequency of the winds that blow from them.

Who does not see the powerful influence such differences in the climatic conditions must exercise on all organized beings, and on vegetation in particular? While, in green Ireland, the myrtle grows in the open air, as in Portugal, without having to dread the cold of winter, the summer sun of this same climate does not succeed in perfectly ripening the plums and the pears, which grow very well in the same latitude on the continent. On the coasts of Cornwall, shrubs as delicate as the laurel or the camelia, are green through the whole year in the gardens, in a latitude at which, in the interior of the continents, trees the most tenacious of life can alone brave the rigor of the winters. But, on the other hand, the mild climate of England cannot ripen the grape, almost unde the same parallel where grow still the delicious wines of the Rhine. At Astracan, on the northern shore of the Caspian, Humboldt says, the grapes and fruits of every kind are as beautiful and luscious as in the Canaries and in Italy; the wines have all the fire of those of the South of Europe, while in the same latitude, at the mouth of the Loire, the vine hardly flourishes at all. And yet, to a summer capable of ripening the southern fruits, succeeds a winter so severe, that the vine-dresser must bury the stock of his vines several feet beneath the earth, if he would not see them killed every year by the cold. Who does not remember that a part of the Russian army, despatched for the conquest of Khovaresmia, perished under the snows, and by the colds of 20° below zero of Fahrenheit, in a country situated under the same parallel as the Azores, where reigns a perpetual spring, and where, in the midst of winter, the vegetation and the flowers display their most brilliant colors? It is there that the camel, the inhabitant of burning deserts, and the reindeer of the frozen regions, meet together, and nature seems to have combined the contrasts of the climate of the poles and of the tropics.

The oceanic climate, considered in the islands truly pelagic, favors the growth of an abundant vegetation, with large and numerous leaves, but little varied. The flora of the oceanic islands, whether from this cause or others pertaining to the mode of dissemination of the plants, is scanty in species. The animal world is still more limited; all the large animals, the lion, the elephant, the rhinoceros, are wanting; the continental islands form an exception, we conceive, because they are much more closely united to the continent than to the ocean.

On man himself the influence of this moist and soft climate makes itself felt, by a relaxation of the tissues, by a want of tonic excitement. The insular Polynesians, as those of Tahiti and others, always exhibit the mild, facile and careless character which seems to be necessarily the result of such a climate.

The continental climate does not give to the vegetation an appearance of such exuberance, but the variety of the soil, the frequency of alternations of plains, of table lands, of mountains, of valleys, of different exposures, secures to it an almost infinite variety of different species and forms. The dryer and warmer air concentrates the vegetable saps, elaborates them better, so to speak, and gives them that strong and aromatic character which the plants of the oceanic islands rarely possess. The animal is more vigorous and larger there, the species more numerous, the types more varied. The lion, the tiger, the elephant, all the kings of the brute creation, have never lived elsewhere than under the sky of the continents, or of the continental islands. Man himself is more animated, more active, more intelligent, endowed with a stronger will; in a word, life is more intense, and raised to a higher degree, by the variety and the movement impressed upon it by the contrasts that form the very esssence of the nature of this climate.

Thus we have two opposite worlds revealed to us, different in their form, their climate, and the organized beings belonging to them. The one, in the main, tends to uniformity, the other to variety. But please to remark, gentlemen, they are not only different from each other, but they stand, moreover, in the relation of superior and inferior.

The terrestrial element has for its portion an infinite variety of the forms of relief, of climate; it is the seat of a more varied life; the birthplace and the habitual abode of all the superior beings, from the vegetable up to man. The ocean has uniformity for its characteristic; it is the domain of the inferior beings, from the polype to the fish and the amphibious animal. Thus we have seen, in the geological development of the surface of our globe, the oceanic element first prevailing, as the less perfect. The oceanic epoch is the embryonic epoch; the insular epoch, analogous to the present oceanic world and its climate, is the second step in the physical life of the globe; the continental epoch, or the present epoch, alone carries it to the highest degree of development.

And yet the ocean much surpasses the continents in extent; it occupies more than two thirds of the surface of the globe. But this even is a sign of inferiority; for mass and number, as we see in all the kingdoms of nature, never belong to the superior being.

At present, gentlemen, we know, in their characters and in their contrast, the continental hemisphere and the oceanic hemisphere; the land and the water. Two different elements are confronted; they cannot remain indifferent; they must act and react, and impart their wealth to each other. We are so much in the habit of seeing these two elements, the dry and the moist, pervading and penetrating one another, that we have some difficulty in figuring to ourselves a state of things wherein the two spheres would be total strangers. We forget that it is to the ocean we owe those beneficent rains, which refresh and vivify all nature; those springs, which quench our thirst; those streams and rivers, which fertilize our valleys and our plains, and serve as highways for the commerce of the nations; those lakes, which spread so many charms over the countries encompassing their borders; we scarcely dream that if the ocean ceased to send to the continents the supply of water necessary to their daily life, the parched and arid earth would soon see all the organized beings that live upon its surface perish in pain and anguish. Desert and death would succeed to life, and at a single stroke the globe would return to the embryonic state of the trilobites, by the extinction of the superior classes of beings.

In fact, all the continental waters come to us from the ocean. If they are fresh and sweet, it is because they have passed through the great laboratory of nature, by a simple process of distillation, which is the first fact that we ought to point out.

The sun, the great awakener of life, the king of nature, shoots his burning rays every day athwart the face of the waters. He causes the invisible vapors to rise, which, lighter than the air itself, unceasingly tend to soar into the atmosphere, filling it and constituting within it another aqueous atmosphere. In their ascending movement, they encounter the colder layers of the higher regions of the atmosphere, which perform the part of coolers. They are condensed in vesicles, that become visible under the form of clouds and fogs. Then, borne along by the winds, whether invisible still, or in the state of clouds, they spread themselves over the continents, and fall in abundant rains upon the ground which they fertilize. All the portion of the atmospheric waters not expended for the benefit of the plants and of the animals, nor carried off anew into the atmosphere by evaporation, returns by the springs and rivers to the ocean, whence it came.

Thus the waters of the ocean, by this ever renewed rotation, spread themselves over the lands; the two elements combine, and become a source of life, far richer and much superior to what either could have produced by its own forces alone.

But we see the earth and the water, the continents and the oceans, touch each other only at their margins A more intimate action upon each other is not possible, except by means of the most mobile of the elements, the atmosphere, performing, in nature, the part of mediator. The winds are the instruments of this important work, the bearers of this wondrous water which renovates unceasingly the face of the main-lands, and sustains their beauty. Unhappy the countries to which they cannot come, still charged with some parts of their precious burthen. The inhabitants of the desert can alone tell us what price we should set upon the smallest portion of this treasure.

To study the distribution of the rains and of the moisture on the surface of the globe, is to study the course of the winds which are their carriers; to this subject, then, we shall turn our attention.

LECTURE VI.

The study of the distribution of the rains supposes that of the winds — Difference of temperature the principal cause of the winds — Theory of the general winds — The winds of the tropical regions — Trade wind of the Pacific Ocean — Trade wind of the Atlantic — The monsoons of the Indian seas — The winds of the temperate regions — Two general currents; the return trade wind, or equatorial wind, and the polar currents — The conflicts of the two, and the variable winds — Lateral displacement of the currents, and their influence upon the temperature, the productions of the soil, and commerce — The law of the rotation of the winds — The atmospheric water falling back in rain — Circumstances favorable to the precipitation of vapors — The rains of the tropical zone — The rains in the region of the monsoons — Annual quantity of the rain water under the tropics — Distribution and annual quantity of the rain in the temperate regions.

Ladies and Gentlemen :—

After having ascertained the characters of the two hemispheres, the oceanic and continental, we have asked ourselves how they acted upon each other, how the moisture of the oceanic climate spread over the continents to fertilize them. We have seen that the atmosphere alone could perform this part of mediator, and that the vapors fly on the wings of the winds to the very heart of the continents. To study the distribution of the rain waters as one of the most essential features of the climate of continents is, as we said, to engage in studying first the movements of the atmosphere, and the general system of the winds. This double study will be the more important, as it is intimately connected with the variations of the temperature, so that it will be almost sufficient to give us an idea of the principal kinds of climates presented by the different countries of the globe.

If we knew only the winds that blow in our temperate regions, we should almost despair of arriving at the knowledge of any law regulating their course. What is more fickle, more capricious, than the winds, which suddenly change their direction, their force and temperature, without apparent cause, and inaccessible to our means of observation? They are the symbols of changeableness itself. But it is not so when we enter upon the equatorial seas, where, from one end of the year to the other, a gentle and regular wind blows from the east to the west with great constancy, and carries slowly and without violence the ships from the coasts of the Old World to those of the New; these are the *trade winds.* We know the astonishment and alarm of the companions of Columbus on noticing these winds, the constant direction of which towards the west seemed to render their return impossible. In the East Indian seas, the winds blow six months from the north-east, and six months from the south-west. These are the *monsoons.* This regularity of the tropical winds indicates the existence of permanent causes, of which it is, perhaps, possible to give some account. At any rate, the phenomenon takes a certain course, annually repeated, of which we ought to take cognizance; for, in case of need, the knowledge of the flow of the atmospheric currents, independently of their causes, may be sufficier for our purpose.

The winds are the consequence of a disturbance of equilibrium in the layers of the atmosphere; and the tendency of their motion is to restore the equilibrium which has been destroyed; as soon as that is accomplished the movement ceases, and everything settles into a calm.

The more we study the causes of these disturbances of the atmospherical equilibrium, and of the winds, the more we see that they are reduced, essentially, almost entirely, to difference of temperature between neighboring places. Here, again, the law of differences is the principle of movement, the condition of life.

One of the chief conditions of the equilibrium of the atmosphere is, that any level layer of the atmosphere should have the same density at all points. If this condition is not fulfilled, the denser portions flow under the less dense, while the lighter rise to the top. Now, this takes place when the different parts of the layer are unequally heated. At the point of greater warmth the air expands, becomes lighter; then, pressed by the neighboring layers, which have remained colder and heavier, it rises into the higher. The result of this process is an ascending current, and lateral currents rushing from all sides towards the spot where the temperature is more elevated. Let us take an example in nature, and see what passes on an island alone in the midst of the ocean.

Let us remember that the land is heated more readily han the sea In proportion as the sun rises above the horizon, the island becomes warmer than the neighboring sea. Their respective atmospheres participate in these unequal temperatures, the fresh air of the sea rushes from all directions under the form of a sea breeze, which makes itself felt along the whole coast, and the warmer and lighter air of the island will ascend into the atmosphere. During the night, it is the reverse. The island loses heat by radiation, and cools quicker than the sea. Its atmosphere, having become heavier, runs into that of the sea, under the form of a land breeze, and this interchange lasts until the temperature, and consequently the density, of the two atmospheres have again become the same. This is the phenomenon observed almost daily on nearly all the seaboards.

What takes place here on a small scale in the space of a day, passes on a great scale between an entire continent and the ocean from one season to another, between the tropical regions and the temperate and polar regions in a permanent manner. Southern Africa is fiercely heated by the rays of a summer sun, while the seas of India and Asia experience the low temperature of the winter. The temperature of the tropics is almost always the same, and constantly higher than that of the rest of the globe. To each of these differences of temperature, unequal in duration and amount, particular atmospheric currents, which are their consequence, correspond; to the difference of temperature between day and night, the diurnal breezes, whether along the coasts or in the interior of the continents, at the foot of the mountains; to the difference of temperature between the extreme seasons, the monsoons, which one might call the season breezes; to the difference of temperature between the tropics and the poles, the trade winds, which are the great annual breeze, and the constancy of which is only the expression of the permanent inequality of the distribution of solar heat between the great atmospherical regions of our globe.

A moment's reflection will enable us to see that these differences of temperature, setting the whole atmosphere in motion, at last connect themselves essentially with the geographical forms of our globe. It is the spherical form which causes the unequal distribution of the rays of the sun, and gives us the great zones of temperature of the astronomical climate, the torrid, temperate, and frozen zones. All the modifications of the solar climate must be referred principally to the geographical forms of the surface, to the distribution and to the relative situation of the continents and the seas.

The general or trade winds are the consequence of the general form of the globe; and their direction, as we shall see by-and-by, is given by its rotatory motion. The monsoons and the breezes depend on the form and the relative situation of the lands and the seas, which govern their intensity and direction. The variable winds are due to the same causes, and to the conflict between the general currents. The primary importance of the geographical forms, which is here revealed at the first glance, will become still more evident in the course of our study.

We shall commence our investigation with the trade

winds, which may be called primitive, of first importance, and which embrace, so to speak, the entire atmosphere. In order to unfold this subject, I shall present the theory generally received by the most eminent meteorologists; that proposed by Halley and Hadley. Not that it is perhaps unassailable in the details, for we encounter many difficulties when we undertake to account by physical laws for the manner in which these great compensations are effected; but the foundations of the hypothesis seem beyond a doubt, and the course of the phenomenon it teaches us to understand is here of the greatest importance.

Let us consider the entire atmosphere as only one of those horizontal layers of air of which we have recently spoken. We see that one of the principal conditions of equilibrium of the molecules does not exist, since the different parts of it are unequally heated. The regions near the equator have a high temperature, and the heat goes on gradually diminishing in proportion as we advance towards the poles. The atmosphere of the tropical zone is more dilated, and consequently lighter, than those of the temperate and polar regions. The height of the barometer at the level of the ocean, which measures the weight of the atmosphere, is in fact less at the equator than in the temperate regions. We have noticed with surprise that the column of mercury, corrected for the effect of the gravity, keeps at a mean of 758 millimetres in the tropics, while it is 761 in the middle latitudes. This difference of three millimetres seems to give the measure of the force which incessantly impels the air of the temperate regions towards the region of the equator.

What is the consequence of this dynamic state of the atmosphere? The denser air of the colder regions presses that of the hot on two sides, the north and the south; the tropical atmosphere rises, and here two lower currents are established, from the poles to the equator, and two superior currents, which conduct the air of the equator towards the poles, to commence again the same rotation. We ought, then, to find, in the northern hemisphere, a general wind coming from the north, and in the southern hemisphere, a wind coming from the south. But the motion of the rotation of the earth from the west to the east exercises an influence upon the direction of these currents, causing them to deviate from their original direction. The speed of rotation, almost nothing in the neighborhood of the poles, becomes greater for any place in proportion to its proximity to the equator. The masses of air rushing towards the equator have then an acquired speed less than that of the regions towards which they are directing themselves. At each step they are obliged to assume a greater rapidity of rotation; but as, in virtue of the law of inertia, a certain time is necessary for this to take place, they find themselves at every step a little behindhand, that is, they are a little further towards the west than would be the case without this circumstance. These successive retardations, accumulating, change little by little the direction of the current from north to south of the northern hemisphere, into a south-west direction; and the direction of the current from south to north of the southern hemisphere, into a north-west current. These two general currents, of north-east and south-east, to call them, according to the usage, by the places whence they come, encountering each other in the tropical zone, combine together, and there results a general current from east to west, the great trade wind. The region where the two currents meet is in a kind of equilibrium, and it is marked by *a zone of calms.*

The same cause makes the upper currents, setting from the equator towards the poles, swerve, but in the opposite way. They arrive successively in the higher latitudes, with a velocity of rotation greater than they find there, and are always a little in advance of the earth's motion in each place; that is, they swerve always more and more to the east. There will then result a current bearing to the north-east, or a south-west wind, in the northern hemisphere, and a current bearing to the south-east, or a north-west wind, in the southern hemisphere.

The general course of the winds would doubtless show itself in all its regularity if the surface of the globe presented only the uniform surface of the oceans. But the presence of the continents and their disposition modify the trade winds in many ways, and make the question very complicated. Let us examine the principal of these modifications, beginning with the trade wind of the tropical regions. In this zone the regularity is greater and the disturbing causes are easier to detect.

The winds of the tropical regions might be reduced to the great equatorial trade wind, blowing regularly from east to west all round the globe, if the continents did not bar its passage and disturb its course at numerous points. The continental lands impede its march, and cut it, so to speak, into several pieces. The trade wind of the Pacific Ocean is arrested by Australia; that of the Indian Ocean by Africa; that of the Atlantic is stopped by America. We shall then rapidly examine the courses of the trade wind in each of these oceans; for it is essentially at the surface of the ocean, where it reigns supreme, that we can learn its true character.

The trade wind of the Pacific begins to make itself felt at a certain distance from the western coasts of America, and blows almost without interruption as far as the coasts of Australia. The north-east current is regular between 2° and 25° north latitude, which may be considered as the southern and northern limits. But in the summer it rises a little further towards the north. It was this constant and gentle wind that carried the first navigator, Magalhaens, whose ship made the voyage round the world, across this vast ocean, and that gave it the name of Pacific, which has been preserved to the present day. It is by this line still, that the Spanish galleons, laden with the gold of the New World, accomplished, during more than two centuries, their peaceful voyages from Acapulco to Manilla, sheltered at once from the tempests and from the attacks of the nations envious of so much wealth. The south-east current is as regular south of the equator, but the limits are less known; it is found as far as the 21° of south latitude.

The region of calms is found in the space comprised between the 2° north latitude and the 2° south, between the two currents at their meeting. Here the ascending

MAP OF THE DISTRIBUTION OF RAIN.

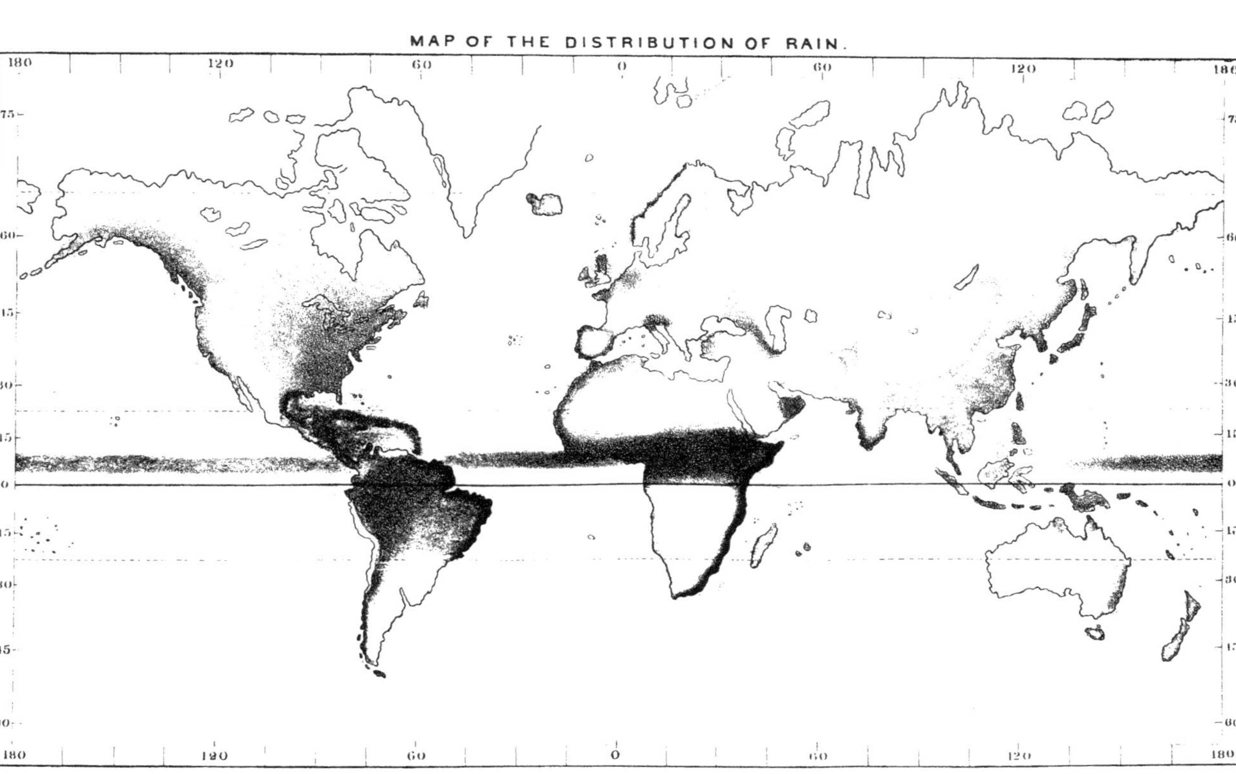

curren seems to neutralize the horizontal; the air is in a sort of factitious equilibrium, that the least accident violently disturbs. Thus, to a dead calm, succeed those sudden tempests, those violent squalls, those whirlwinds, those tornadoes, as the Spaniards call them, which are the terror of navigators. Thunder storms, accompanied by showers, are of almost daily occurrence.

The trade wind of the Atlantic is already modified by the position of this ocean lying between continents nearer to each other. It is, as it were, transported bodily several degrees towards the north. The northern limit of the north-east current is precisely fixed by the numerous navigators who traverse these seas; it commences between 28° and 30° north latitude. Its southern limit is about 8° north latitude. The region of calms occupies, on the average, the space comprised between the 3° and 8° of north latitude; but its position varies with the seasons; in August it extends from 3° to 13° north latitude; in February, from 1° to 6° north latitude. The south-east current always blows, then, beyond the equator to the north.

Humboldt attributes, apparently with reason, this anomaly, on the one hand, to the direction of the coasts of South America, which favors the extension of the south-east trade wind, and of the warm waters of the great equatorial current towards the north; and, on the other, to the cooling influence of the high mountains of the continent, in the regions of the equator. The first of these causes tends to heat the sea of the Antilles; the second, to lower the temperature of the southern continent. The result of this difference must be to determine a current of air from the south, removing the limit of the north-east trade wind further north. The thermal equator, or the line of the greatest mean heat, passes, in fact, through the south of the sea of the Antilles.

The existence of the upper trade wind, coming from the west, or of the return trade wind, which has often been doubted, seems to be proved in this ocean by two facts, often cited and very conclusive. The volcano of the island of St. Vincent, belonging to the lesser Antilles, in one of its eruptions hurled a column of volcanic cinders to a great height in the atmosphere; the inhabitants of the Barbadoes, situated east of St. Vincent, saw, with astonishment, the cinders falling in abundance upon their island. The 25th of February, 1835, the volcano of Cosiguina, in Guatemala, threw into the air such a quantity of cinders, that the light of the sun was darkened during five days; a few days after, they were seen to cover the streets of Kingston, in Jamaica, situated north-east of Guatemala. In these two cases it is evident that the cinders had reached the region of the upper trade wind, and had been carried by it from west to east, in the opposite direction to the lower trade wind. At the summit of the Peak of Teneriffe, most travellers have found a west wind, even when the north-east trade wind prevailed on the seaboard.

The winds of the Indian Ocean experience still greater perturbations than those of the other two oceans of the tropics. If I have elsewhere called the Pacific the most *oceanic* of the oceans, the Atlantic the most *maritime*, I will call the Indian Ocean the most *mediterranean*. It is, in reality, only a half ocean, a great gulf, surrounded on the sides by huge continental masses; the mighty Asia, with its peninsulas and its table lands, on the north, Africa on the west, Australia on the east. Asia prevents the oceanic trade wind of the north-east from arriving there, and the influence of the lands and of the vast plateaus remains greatly preponderant. Thus the movements of the atmosphere depend upon the unequal heating of the neighboring continents during the extreme seasons of summer and winter, which are opposite in the continents situated in the north and in the south. The eastern trade wind in this way changes into a sort of double semi-annual breeze, blowing regularly six months in one direction, and six months in another; this is called *monsoon*, from the Arabic word *moussin*, signifying season. It will be easy to understand this effect, if you call to mind what we have said of the land and sea breezes, that spring up on the islands and along the sea-shores.

While Africa, south of the equator, receives the vertical rays of the southern summer sun in December, January, and February, Southern Asia on the north of the equator, and the neighboring seas, are feeling the low temperatures of winter. The air rushes in from the colder regions of the Indies and of Upper Asia towards the warmer regions of Southern Africa, and the trade wind is transformed into a north-easter, which blows as long as this difference of temperature lasts. It is for India the winter or north-east monsoon. The reverse takes place when India and Asia are heated by the burning sun of the northern summer, and when Africa is cooled by the southern winter. The air blows towards the places where the temperature is more elevated; it is for India the summer or south-west monsoon.

Hence, in place of a constant current setting from east to west, the relative position of the lands, combined with the action of the earth's rotation, gives occasion to two periodical winds; the monsoon of the south-west, blowing from April to October, during the northern summer, and the north-east monsoon, blowing from October to April, during the southern summer. In the southern part of the Indian Ocean, which is not under the influence of the lands, the south-east trade wind blows quite regularly through the whole year.

The transition from one monsoon to another, depending upon the course of the sun, does not occur at the same period in places situated under different latitudes; but the approach of this critical season is always heralded by variable winds, succeeded by intervals of calm, and by furious tempests and whirlwinds, proving a general disturbance of the atmosphere.

The phenomenon of the monsoon, or the change of winds according to the seasons, takes place in like manner between the Indies and New Holland. But it is less regular and less marked than the Indo-African system we have just described. The seas of Southern China and the great archipelago of Sunda and of the Moluccas, by their position feeling at the same time the influence of the trade wind of the great ocean, and of the double system of the monsoons of the Indies and Australia, it is easily conceived that we must seek in this circumstance the cause of the tempests and typhoons which desolate this sea more than any other upon the surface of the globe.

We see that the great trade wind does not exhibit its normal manner, except in the Pacific Ocean, far from the land. It is driven towards the north in the valley of the Atlantic, or is entirely broken up in the Indian Ocean. The influence of the lands cannot be mistaken.

Let us pass to the winds of the temperate regions and of the middle latitudes.

Here, as we have said, the regularity disappears by degrees; the secondary influence assumes more importance still; it is the theatre of the incessant conflict between the polar winds and those of the tropics. They blow alternately, without any well established rule, and pass, often abruptly and without transition, from one point of the horizon to another. If the equatorial regions are those of the *constant and periodical* winds, the temperate regions are those of the *variable winds.*

Nevertheless, when we compare the number of times the winds blow from each quarter of the horizon during the course of a year, we discover that in the northern hemisphere two directions tend to prevail over all the others, and those are the winds from the west and south-west, and from the east and south-east. It is known that in the northern Atlantic the west winds prevail to such a degree that the average passage of the packet ships from America to Europe is only from twenty to twenty-three days, while from Europe to America it is from thirty-five to forty.

It is generally agreed to consider these winds from the south-west as having their origin in the return of the air of the tropics. The upper trade wind cools in the high regions of the atmosphere, and descends again to the surface, reaching it about 30° N. latitude, or even still further north, during the summer. In winter, the limit where the winds from the north-east and south-west change place, is marked by variable winds and calms, which the navigators, coming from the north, ordinarily encounter before entering the region of the trade winds. The long Atlantic valley is the grand route of the winds of the equator; they spread themselves there without obstacle, beyond the influence of the lands; and the line of the coasts of America, as the direction of the ocean itself, coincides with that impressed on them by the earth's rotation. They advance as far as the high latitudes of Norway, near the polar regions, and bathe all the western coasts of Europe in their soft and humid air.

The northward inflection of the lines of mean equal heat, or of the isothermal lines, which you see traced on the map before you, (See plate I.,) shows us at a glance the considerable influence of the winds upon the temperature of the Atlantic, and of the western coasts of the Old World. It is such, that in Europe some of the cereal grains grow even at Cape North in the latitude of Boothia Felix, about the coldest point ascertained on the globe; and that the brilliant cities of Stockholm and of Petersburg flourish under the parallel of the regions of eternal ice in Northern Labrador.

What I have just said of the return of the trade wind in the Atlantic, is true again for the Pacific Ocean. The winds of the west and south-west prevail in the middle latitudes; they strike the western coasts of North America and carry thither the soft temperature belonging to them. Sitka, in Russian America, at 57° N. lat., has the same average temperature with the shores of Lake Ontario, 44° N. lat., but much milder winters; the valley of the Columbia, in Oregon, already displays the most verdant prairies; while, under the same latitude, Lake Superior presents only snow and ice, and the whole desolate aspect of an arctic region.

It is, then, to the normal direction of the return trade wind that we must refer the well-known phenomenon of the higher temperature of the western shores of the continents of the two worlds, compared to that of their eastern sea-boards. But, for the same reason, this difference, though very great in the high latitudes, disappears by degrees as we approach the tropics.

But the air of the polar countries, tending continually to flow towards the warmer regions, gives birth to currents, the normal direction of which is from north-east to south-west, from the cause we have already explained. These north-easterly currents follow, by preference, the path of the continents, as the currents of the equator follow that of the ocean. They have the cold temperature of the places whence they come, and unless high mountains interpose an obstacle, they refresh the continental regions for a great distance. Cast a glance upon this map of Europe, where the lines of equal mean temperature are traced, (See plate I.,) and you will see them strongly bending towards the south opposite to the broad passage opened to the polar winds between the Caucasus and the mountains of Transylvania; that is, all the borders of the Black Sea, the northern coast of Asia Minor, the eastern coast of Greece, owe to them a lower temperature than that found at the same latitude in the neighboring countries, sheltered against the attacks of this icy Boreas, by high chains of mountains.

The polar winds play equally a very important part in the climate of North America. No other continent offers them a more open path from one end to the other of its extent. From the borders of the Frozen Ocean, to the subtropical regions as far as the Gulf of Mexico, no chain of mountains opposes their unobstructed sweep; for they are all directed from the north to the south. Almost no spot is sheltered from their sudden and cold attacks. Nevertheless, owing to the disposition of the Atlantic coasts, retreating on the north-east, and the south-west direction taken by all the currents of the north, the west and the south-west bear the first shock. These polar winds, it seems, strike obliquely against the mass of the Rocky Mountains, run along their slopes, and, being guided and reflected by this high chain, descend under the form of a north-west wind into the valley of the Mississippi, accompanied by cold and storms, and advance towards the Atlantic coast. In this route they encounter the return trade wind, the south-west current, which they take in flank; and I incline to think, that to this conflict are owing some of those tempestuous storms, revolving from east to west, the course of which has been so well described by Mr. Redfield. Others, as the same learned man has triumphantly demonstrated, have their origin in the tropical seas.

If this conflict of the two currents of air often commences at the south of the continent, and seems to advance towards the north, it is by reason of its form and of the disposition of its shores, approaching each other towards the south and greatly diverging towards the north. The western coast and the Rocky Mountains trend thus to the north-west; the Atlantic coast to the north-east. Now, supposing the mass of air, turned aside by the Rocky Mountains, to advance from west to east on a line nearly parallel to this chain, it first strikes the Atlantic region in the south, then successively reaches points more and more towards the north.

This conflict of polar and equatorial winds, opposite in character and direction, gives to our climate one of its most characteristic features, that changeableness, that extreme variety of temperature, of dryness, and of moisture, of fair weather and of foul; that uncertainty of the seasons which always keeps the merchant and the farmer in anxious suspense, between the hope of a good harvest and the fear of a dearth.

Not only are the variations in the same year considerable, but they are still more so from one year to another. The system of these currents oscillates from east to west, and changes place. The polar winds will prevail in a country, and will endanger the crops by the prolonged dryness of their atmosphere; while further east or west the trade wind will spread fertility by its beneficent rains. Or the opposite: the south winds acquire such a preponderance, that the harvests perish by the moisture, while at a somewhat greater distance, on the limit of the same wind, nature lavishes all her treasures upon the laborer. It has been remarked that a mild winter in Europe corresponds frequently to a severe winter in America and Asia; while the mildness of the winter in America affords a presumption of a colder winter on the other side of the Atlantic. The years 1816 and 1817 were marked, as is known, in the history of Europe, by a general famine and distress. The wet was such that the harvests failed entirely. But the south-west wind, which blew without cessation over the western part of the continent, and drenched it in its vapors, did not extend beyond Poland; and it was the South of Russia whose corn supported famished Europe for many long months. Then was revealed the commercial importance of these countries, hitherto unknown, and constantly increasing since. Who does not still remember the immense impulse given to the commerce between Europe and America, by the drought of 1846, which damaged the corn crop in Europe, while America had an abundant harvest? These examples alone tell us the important part played in the life of the nations by those variations of the atmospheric currents belonging to our temperate countries.

In the middle latitudes of the northern hemisphere, there are, properly speaking, only two normal winds, that of the north-east and that of the south-west. The winds blowing in other directions, are local winds, or transition winds, from one of the general currents to the other. Professor Dove has shown, that, in Europe at least, these winds succeed each other in an order always the same, which he has called the law of rotation of the winds. This will be easily understood if we remember that in advancing along their course, the south-west wind tends always to become more west, and the north-east more and more east; we shall see that the result of this disposition ought to be, wherever they meet each other and change places, a rotation from west by north to east, and from east by south to west. In the place of the conflict of the two currents, the wind will then blow successively from these different regions, and in this order, until it is established in the direction of that one of the currents which has overpowered the other. But no one of these transition winds blows for any great length of time. In the southern hemisphere the order of succession is the reverse.

The course of the winds being explained, it will be easy to understand the distribution of the rains on the surface of the globe.

The winds sweep in all directions, as we have just seen; they carry with them into the places where they go, the temperature and the moisture of the places whence they come. A sea breeze will be always moist and relatively temperate; a land wind, dry and extreme, whether in cold or in heat. The first, ordinarily, is the herald of rain; the second, fair weather. It is the atmosphere that brings into connection the most distant countries of the globe, with regard to temperature and humidity, and softens all the differences by blending opposite and extreme characters.

We have seen how the atmosphere is charged with the vapors of the ocean, but we have not stated how it happens that these vapors are condensed anew, to fall again in rain. This depends chiefly on a property of the air, of which we must say a word.

A determinate volume of air, a cubic foot, for example, at a given temperature, has the property of receiving a certain quantity of vapor, of water in an invisible state, or, as we call it, *humidity*. When it contains all the humidity it is capable of receiving, it is said to be *saturated*. If you increase the temperature, it will be able to hold more; if, on the contrary, you lower the temperature, you diminish its capacity for vapor, and, in the given case, a part of the vapor would be condensed and deposited in small drops of rain along the outside of the vessel. The moist air here is like a sponge filled with water; reduce its volume by pressure, there will run out a certain quantity of water; in the air laden with moisture the diminution of the temperature takes the place of pressure.

We can easily conceive the application of this principle in meteorology.

A warm and moist wind, the south-west of the Atlantic, for example, setting from the tropics, comes in contact with the colder air of the temperate regions; its temperature is lowered; it can no longer contain as great a quantity of vapor. A portion of its humidity is immediately condensed into clouds, then falls in rain.

Or the opposite; a wind charged with clouds arrives in a warmer and dryer air; comes, for example, from the Mediterranean to the Sahara, as is the case during three fourths of the year; the burning air of the desert, having a much greater capacity for vapor, dissipates instantly all these clouds, that break up, vanish, and disappoint the excited expectation of the traveller, who hoped for refreshing rains.

Do the moist winds encounter an elevated obstacle, a high chain of mountains, a plateau? Forced to ascend their slopes, high into the atmosphere, they find there a colder air, which condenses their vapors, and the rain flows down along the sides. The wind passes over to the other side of the chain; it arrives dry and cold, deprived of all its moisture, without clouds. The same wind thus brings rain on one side, and fair weather on the other. This is what happens every day on the two sides of the Scandinavian mountains.

It is even possible that an ascending current, if very violent, may hurry the abundant vapors of the lower layers to the more elevated layers of the atmosphere. The vapors are afterwards condensed there, and fall back in torrents of rain. Such at least is the explanation Humboldt gives of the rains of the tropics.

Aided by these preliminary remarks, we are enabled to account for the general phenomena regarding the distribution of rains, which I desire to explain to you. We will therefore devote the remainder of our time this evening to following out the general march of this phenomenon in the tropical regions and the temperate zones.

The temperature, the winds, and the rain, having an intimate connection each with the others, and playing alternately the part of cause and effect the earth, in the point of view now under consideration, is divided, as in the point of view of temperature and winds, into two great zones — the one, that of periodical rains, or of the tropical regions; — the other, that of continuous rains, or of the temperate regions.

In the equatorial regions, where the course of temperatures and winds is regular, that of the rains is equally so; and instead of seasons of temperature, which are there unknown, the inhabitants draw the distinguishing line between the dry and the rainy season.

Whenever the trade wind blows with its wonted regularity, the sky preserves a constant serenity, and a deep azure blue, especially when the sun is in the opposite hemisphere; the air is dry, and the atmosphere cloudless. But in proportion as the sun approaches the zenith, the trade wind grows irregular, the sky assumes a whitish tint, it becomes overcast, clouds appear, sudden showers, accompanied with fierce storms, ensue. They occur more and more frequently, and turn at length into floods of rain, inundating the earth with torrents of water. The air is at this time so damp that the inhabitants are in an incessant vapor bath. The heat is heavy and stifling; the body becomes dull and enervated; this is the period of those endemical fevers that destroy so great a number of the settlers who have come from the temperate zones. But vegetation puts on a new freshness and vigor; the desert itself becomes animated, and is overspread for a few months with enchanting verdure, which furnishes pasture to thousands of animals. Nevertheless, ere long, the sun, in his annual progress, advances to pour down his vertical rays upon other places; the rains diminish, the atmosphere becomes once more serene, the trade wind resumes its regularity, and the heaven shuts its windows once again until the following season.

Such is the normal course of the tropical rains. They fa... everywhere during the passage of the sun through the zenith. The heat is then so violent that the ascending current neutralizes the horizontal trade wind. It hurries the vapors to the heights of the atmosphere, and the upper limit of the trade wind, where they are condensed and fall back in a deluge of rain. Now, as the sun passes and repasses from one tropic to the other, it follows that there is, in most intermediate places, a twofold rainy season, the two periods of rain being more or less closely connected in point of time.

In India the course of the rain is not so regular; it depends entirely on the monsoons. The western coast of Deccan, the coast of Malabar, has the season of the rains during the monsoon of the south-west, which brings thither the vapors of the ocean; that is, during the northern summer. It has the dry season during the monsoon of the north-east. During the winter, the monsoon of the south-west ascends the slopes of the western Ghauts, and causes, in the heights, violent storms and very abundant rains. Along the coast of Coromandel, on the contrary, it is the north-east monsoon which conducts the rains, with the vapors of the Sea of Bengal, and the south-west monsoon brings the dryness. These two coasts of the peninsula have then their seasons reversed. One has the dry weather when the other has rain, and reciprocally. The table land of Deccan partakes of the two characters; the fall of water is more variable, and there are often two periods of abundant rains.

We see here that the relative position of the lands and the seas regulates the seasons.

The quantity of water that falls from the atmosphere in the tropical regions during some months is enormous, if we compare it with that which we are accustomed to see wetting the soil of our own countries. It has been calculated, that, on the average, there fall annually in the tropics of the Old World 77 inches of water, and 115 in tropical America. The mean for the equatorial region would be 96 inches.

But the annual quantity of rain received in some localities, and under the influence of certain circumstances, is sometimes much more considerable. At Paramaribo, in Dutch Guiana, it falls to the amount of 229 inches of water, or 19 feet. At St. Louis de Maranhao, in Brazil, 276 inches have been received, or 23 feet. But the greatest quantity ever observed is that of Mahabaleshwar, in the western Ghauts, south of Bombay, at the height of 4,200 feet; it rises to the enormous number of 302 inches. A layer of 25 feet of water would have been formed by the rain waters, if they had not gradually run off.

These results are the more astonishing, as all this water falls in the space of only a few months, and, so to speak, at once. It has been seen to fall at Cayenne 21 inches in a single day. This is nearly as much as falls during the whole year in the northern latitudes. This is the reason why, notwithstanding the abundance of the rains, the number of clear days is much more considerable than in our climates. Even during the rainy season, the sun shows himself nearly every day, and many days pass without a single drop of water falling from the atmosphere.

We may conceive the prodigious effect such violent showers must produce upon the rivers. Who does not now understand the secret of the overflowings of the Nile, once so mysterious, which are due to the circumstance that the region of its sources receives the tropical rains?

Floods of forty feet rise and upwards are frequent at this season in the great rivers of South America; the llanos of the Orinoco are changed into an inland sea. The Amazon inundates to a vast distance the plains it flows through. The Paraguay forms lagoons, which, like those of Xarayes, are more than three hundred miles in length, and ooze away during the dry season.

The quantity of water contained in the tropical atmosphere in the condition of transparent gas, is always considerable. It is in proportion to the heat, which, being always very great, augments its capacity to a very high degree. Even under the most serene sky, the air is still abundantly provided with it. It is this invisible water which, being absorbed by the plants, and taken up by their large leaves, produces the vigorous vegetation, and causes the eternal verdure that fills us with astonishment, under a sky devoid of rain, and cloudless during more than half the year; while in our climates, from the failure of rain for a few weeks only, we see all verdure languish, and all the flowers perish for the lack of moisture.

The distribution of rains in the temperate regions offers a perfect contrast to that of the tropics. Here, throughout the whole year, the earth is watered by the rains of heaven, although sometimes irregularly. But these are variable, as are the winds and the temperature, and secondary circumstances have much influence over them.

The further we recede from the tropics, the more do we find that this periodical character disappears. But we have few established facts as to the mode in which the transition is made from one region to another. North of the tropics we find winter rains, which doubtless are caused by the meeting of the upper trade winds with the north-easters. The strife of these gives birth to heavy rain storms. It is so at Madeira and Lisbon. Yet further north, Italy, and some portion of the Mediterranean, have spring and autumnal rains, which Dove attributes to the transit of the south-west trade wind before and after the solstice. In Germany, according to the same authority, the same cause produces frequent rains at the period of the solstice, or summer rains, denoting the highest point attained by the trade wind in those latitudes at the greatest declination of the sun towards the north.

But it must be admitted that the general character of the rains of those regions, their periods, and their frequency, appear especially to depend on a thousand geographical features which influence them greatly.

The quantity of water held by the atmosphere of the temperate regions is much smaller than that in the air of the tropics. The vegetation, therefore, cannot endure the want of rain for any length of time, as I have observed, and the quantity of rain water falling in them is also greatly inferior.

The mean is 34 inches in the Old World, and 39 in temperate America, or 35 for the whole zone. There are causes, however, to be pointed out hereafter, which produce the fall of twice and even thrice that quantity at certain points. The number of fair days is also far smaller. But if these fruitful showers are not granted to us with the same prodigality as in the tropical regions, they are, at least, better distributed throughout the year, in a manner more equal, more economical, and more advantageous to vegetation and the requirements of all organized beings.

LECTURE VII.

Modifications of the general laws of distribution of the rains — Decrease of the quantity of rain waters and of rainy days, from the seaboard towards the inlands — Numerous exceptions and their causes — Influence of the mountains and the table lands in the two worlds — Distribution of rain in South America; in North America; in Africa; in Europe; in Asia; in Australia — Special hygrometrical character of each continent — Difference between the Old and the New World, corresponding to the nature of their relief — Mixture of the continental and the oceanic element — Influence on organized beings — Superiority of the zone of contact, or the maritime zone.

Ladies and Gentlemen: —

The investigation we attempted to make in the last lecture, has convinced us of the intimate connection existing between the temperatures and the winds, and between both and the distribution of rain over the surface of the earth. In this last point of view we have recognized the existence of a zone of *periodical* rains, corresponding to the torrid regions of the equator, wherein the rains fall in abundance, and within the space of a few months; and of two zones of *continuous* rains corresponding to the temperate and cold regions, in which they fall in smaller quantity, and are more uniformly distributed through the entire course of the year. It remains for us, this evening, to give some account of the numerous modifications these general laws are made to undergo, by the extent of the continents, the forms of their relief, and their position relatively to the general winds which are the dispensers of the rain waters.

The map before us, on which it has been attempted to express, by deeper or lighter tints, the relative abundance of the rain that falls in each region of the globe, indicates these zones in a very clear manner; it will serve further to illustrate what remains to be said on this subject. (See plate IV.)

The winds of the ocean striking the coasts of the continents, and moistening them with their waters, penetrate equally into the interior, transport thither the vapors with which they are charged, and spread life and freshness on their path. But in proportion as they advance on their continental journey, they become more and more scant and sparing of these beneficent waters; their provision is exhausted, and if the way is too long, if the continent is too extended, they arrive at its centre, as arid and parched as a land wind.

This first result appears so natural, that it seems almost useless to exhibit it by figures. Nevertheless, we will let direct observation speak, that the fact may not rest upon assertion alone. Here is the quantity of rain water received annually in the different parts of the same continent, more or less remote from the seaboard. I add also the number of rainy days, to complete these observations. As far as possible, I choose countries situated under similar latitudes, in order to render them capable of a more rigorous comparison in this point of view; for, otherwise, the quantity of rain water diminishing in proportion to the distance from the torrid regions of the equator, it would be easy to attribute incorrectly to the distance from the seas a difference that might be only the effect of a position more or less towards the north.

The mean quantity of rain received during a year, and the number of rainy days, are as follows, in the countries situated between 45° and 50° N. lat. of the Old World:

	Depth of Rain in inches.	Number of Rainy Days.
British Islands,	32	156
Western France,	25	152
Eastern France,	22	147
Central and North Germany,	20	150
Hungary,	17	111
Eastern Russia, Kasan,	14	90
Siberia, Yakoutsk,	?	60

We see that, in leaving the coasts for the interior of the continents, there is a gradual diminution of the quantity of rain and of rainy days. If we penetrate to the centre of the vast continent of Asia, we find the dryness there almost absolute—a desert.

In North America, the observations are as yet so few and so recent, that it is impossible to deduce from them very exact averages. Besides, as we shall soon see, this continent being exposed at the same time to the winds of the Atlantic on the east, and to those of the Gulf of Mexico on the south, receives rain waters from both directions. This is especially true of the middle region, situated west of the Alleghanies. In this way the decrease, owing to the distance from the Atlantic, is disguised by the additional rain water brought thither by the winds of the Gulf of Mexico. These various circumstances tend in a singular degree to render the distribution of the rains more uniform in this part of the continent. Nevertheless, the following numbers seem to indicate that the influence of the continental position is not annihilated.

The annual quantity of rain water between 41° and 43° north lat. is, at

	Depth of Rain in inches.
Cambridge, Mass., ...	38
Western Reserve College, Ohio, ...	36
Fort Crawford, Wisconsin, ...	30

Again, between lat. 38° and 40° north.

	Depth of Rain in inches.
Philadelphia, Pa., and Lambertville, N. J., . .	45
Marietta, Ohio, ...	41
St. Louis, Missouri, ...	32

We may say, then, that, in general, a country is the better watered, the nearer it is to the seaboard; and, from moist and verdant Ireland to the desert of Gobi, we find all possible gradations between the extremes of moisture and aridity.

This indubitable general law, however, undergoes numerous modifications, which infinitely diversify the nature of the climates in regard to their wetness or drought, causing the most surprising anomalies.

On the shore of the Caribbean Sea, on the coast of Venezuela, is situated the city of Cumana, which has become celebrated in the annals of science by the researches made there by Humboldt. That city, in the midst of the regions of the tropics, where the rains are so abundant, in spite of its maritime position, receives only 8 inches of water, while very near it, a little further south-east, in Guyana, there is a fall of more than 200 inches.

In this same South America, so plentifully watered, we see on the opposite side, south-west of the Andes of Bolivia, a long and narrow band destitute of rain, stretching several hundred miles along the coast; it is the desert of Atacama. Not a drop of water comes to refresh this thirsty land, though lying upon the sea-coast, and under the same latitude as the plains of Upper Paraguay, which is inundated with rain.

The plateaus of Upper California are nearer the sea than the centre of the valley of the Mississippi, and, nevertheless, they are dry and parched, while the latter is fertilized by copious rains.

Here are causes, then, which disturb the general law, or rather which modify it in favor of variety of climates; these causes are the forms of relief of the soil, the mountain chains and the plateaus, and their disposition relative to the damp winds.

A wind loaded with vapor and clouds may pass over vast continental plains, without dissolving into rain, because the temperature in a plain may remain the same through long spaces, or even be higher than that of the sea wind crossing it. There is, then, no cause of condensation of the vapors. We have an example of this in the Etesian winds, which bear the vapors of the Mediterranean into Sahara. They have no sooner passed the threshold of the desert, than the dry and burnt air, as we have already said, dissipates even the smallest cloud.

But it is not the same when the moist winds meet elevated objects, chains of mountains, and high table lands, in their transit. Forced to ascend along their sides, they are uplifted into the colder regions of the atmosphere; they feel the pressure of the air, which is less there, and the expansion of the gases composing them further increases the cooling; the air loses its capacity for holding the same quantity of vapors as before. The latter are condensed into clouds, which crown the summits of the mountains, and trail along their sides; and they melt soon into abundant rains. If the sea wind passes the chain, it descends on the opposite side, dry and cold; it has lost all its marine character.

The mountain chains are, then, the great condensers, placed by nature here and there along the continents, to rob the winds of their treasures, to serve as reservoirs for the rain waters, and to distribute them afterwards, as they are needed, over the surrounding plains. Their wet and cloudy summits seem to be untiringly occupied with this important work. From their sides flow numberless torrents and rivers, carrying in all directions wealth and life. Every system of mountains becomes the centre of a system of irrigation, of water courses, which gives to its neighborhood a value of primary importance.

This power of condensation is expressed by the fact, that in the heights of the mountains there falls more water than on the slopes, and at their foot there falls more than in the neighboring plains. Further, the side of the chain exposed to the sea winds receives a quantity of rain much beyond that which falls on the opposite side; so that the great systems of mountains not only divide the spaces, but separate different, and often oppo-site, climates.

The examples of this action of mountain chains on the condensation of the rains, are numerous in nature. I have only an embarrassment of choice. Nevertheless, I am compelled to borrow them from the Old World, because the exact observations I need are there more numerous.

The Alps form a vast semicircle on the north of Italy, wherein the warm and moist winds of the south-west, coming from the Mediterranean and the ocean, pour themselves as into a funnel. Before passing this lofty barrier and the snow-capped summits, these winds lose their vapors, which fall in copious rain on all the southern slope of the chain. While 36 inches of water fall in the plains of Lombardy, there falls an average of 58 inches at the very foot of the Alps. In the north-east corner, forming an angle, where the vapors accumulate at Tolmezzo, in the valley of the Tagliamento, a quantity of 90 inches annually is received, which reminds us of that of the tropical regions. Now, this number is a very constant one, for it is the average of twenty-two years' observations. The northern foot of the Alps has only 35 inches.

The Apennines repeat almost the same phenomenon. They form an arch, the convexity of which is marked by the curve of the Gulf of Genoa and the valley of the Arno. The summits, which rise from 4,000 to 6,000 feet, arrest the winds of the sea, and there fall at their southern foot 64 inches of water, while only 26 inches fall on the northern descent, in the plains south of the Po. The same relation exists further south, between the western and eastern slopes of the same chain; on the former it rains 35 inches of water; on the latter only 27.

We have already quoted Scandinavia as giving one of the most striking examples of this kind of phenomena. The elevation and the length of that chain, its lofty frozen table lands, which a long day's journey is hardly sufficient to cross, are an insurmountable barrier to the vapors brought thither on the Norwegian coast by the south-west wind from the Atlantic. They are condensed almost entirely upon the shores incessantly plunged in drizzling fogs. At Bergen, a day of sunshine is a rarity, in the midst of almost constant rains that darken the atmosphere. Thus we have there a fall of 82 inches of water, — an enormous quantity, especially for such high latitudes. All the western coast receives nearly as much, and owes to the temperature of this wind, and to the caloric disengaged by so active a condensation of vapors, the remarkably soft and equable climate which distinguishes it. On the southern coast, and in Sweden, there fall only 21 inches of water, and the same south-west wind brings thither clear weather and cold. The same wind carries rain on one side, and fair weather on the other.

In the East Indies, we encounter the majestic chain

of the Himalaya, the most massive and lofty on the globe.

The winds of the tropics, passing over the plains of the Ganges, reach it, water the southern slopes, fertilize the inland valleys, and support the most wonderful verdure, up to the limit of eternal snows. But beyond, the table lands of the region of the sacred lakes and of Katchi and Tangout, indicate by their drought that they are deprived of this beneficent influence. Katmandoo, at a third of the height, has 51 inches of rain; Delhi, in the plains of the Ganges, has only 23.

At the north-east angle of the Indo-Persian Sea, the south-west trade wind accumulates its vapors on the flanks of the Ghauts. The effect of this chain, which, however, has no great elevation, is such, that, after the following examples, we shall be able to dispense with any more. At Bombay, on the west coast, the rain falls 80 inches; 302 have been received at Mahabaleshwar, on the mountains, at an elevation of 4,200 feet, as we have already said; this quantity is reduced to 26 inches on the other side of the chain, at Darwar, on the table land of Deccan.

But we have said that the plateaus also have a marked effect upon the distribution of the rain waters. Their borders act as the mountains, and their surface, heated more than the layers of air of the same level, absorb the little vapor which ascends to this height, without condensing it; their extent, finally, and their elevation, tend to impede the access of the oceanic vapors, and to increase the drought. These differences are already marked in plateaus so little elevated as Spain, whose central plains are from 2,000 to 2,500 feet above the sea. While the south-west coast of Portugal, — Lisbon for example, — is watered with 27 inches of rain, the border of the table land has only 11 inches; and soon, quitting the verdant region of the seaboard, we ascend the arid plains of Estramadura, of La Mancha, and of Castile, at the centre of which, Madrid receives not more than 10 inches of rain water. No other place in Europe is so badly provided in this respect. And, nevertheless, side by side with this minimum of rain, we find the greatest quantity that has ever been made out on this continent. At the western foot of the Sierra d'Estrella, which advances, like a spur, very far towards the coast, in the valley of the Mondego, there has been received, it is said, at Coimbra, the enormous quantity of 225 inches of water. An error has been suspected in this measure, taken in 1816 and 1817. Schouw has reduced it to 135 inches; Kæmtz, to 118; adopting the last number, there is still a difference of more than 100 inches from Madrid, situated under very nearly the same latitude, and on the same peninsula.

If it is so with the table lands of the third order, as that of Spain, what will be the case with those enormous masses which form the body of eastern and western Asia?

The fringe of snowy mountains surrounding them, their distance from the oceans, the extent of their surfaces, their elevation in the atmosphere, — all these causes conspire to give them that character of aridity which renders them almost an unbroken desert.

The plateaus of southern Africa, those of Mexico and of California, compared with the neighboring countries, have equally an indisputable character of aridity. At Vera Cruz, for example, there fall 62 inches of water, while in Mexico, and on the coast of the Pacific, the quantity seems to be considerably reduced.

If the influence of mountain chains and table lands is so considerable in all the particular cases which we have just examined, it ought to manifest itself on a grand scale and in a certain connection, for each continent in particular, and for each of the two worlds. We have previously ascertained a general law of the distribution of the reliefs; there should be here a reflection of this law; and its importance should be revealed in the distribution of the pluvial waters, and of the climate. We proceed, then, to seek an explanation of the effect that must be produced upon each continent by the particular disposition of its chains of mountains, of its plateaus and plains, relatively to the maritime winds, bringing them the rains and tempering their climate.

Let us begin with the New World, the structure of which is more simple and easy to comprehend.

The fundamental features of the structure of America, I repeat here, are the long and lofty barrier of the Andes, of the Rocky Mountains, extending almost from one pole to the other, along the western coast of the two continents; then, on the east, vast plains, interspersed with some mountain ranges of slight elevation. Let us see what is the effect of this disposition on the climate of both these continents.

In South America, the principal body of which is situated under the sky of the tropics, this disposition secures to the continent a copious supply of moisture. The plains of the east are open to the trade wind of the Atlantic, which sweeps over them unobstructed, and bears thither unceasingly the vapors of the ocean. The secondary chains of Brazil and of the Guyanas, from 5,000 to 7,000, do not rise high enough into the atmosphere to arrest it; the only effect they have, is to augment the falling showers, and to supply a more complete irrigation. The Orinoco, and the lower tributaries of the Amazon, the Tocantins, the San Francisco, and many others which flow from these two systems, are there to tell us. But it is not the same with the Andes. This chain, whose crests and summits lift themselves everywhere into the region of perpetual snows, forms, by its elevation and continuity, an invincible obstacle to all the moist winds of the east. The vapors, having traversed the plateaus of Eastern Brazil, without lingering there long, accumulate and condense, and flow down their eastern slopes. All this zone at the foot of the Andes is one of the best watered in the globe. Thus we see issuing from hence those immense streams; the Marañon, the king of the rivers of the earth, and all its tributaries, the Ucayale, the Rio Purus, the Madera, and many others, to which nothing is wanting but to flow through civilized countries, in order to rival in importance the Nile, the Ganges, and the Mississippi.

But on the other side of the Andes all is changed. Neither the trade wind nor its vapors arrive at the western coast. Scarcely do the table lands of Peru and of Bolivia receive from them the latter benefits, by the storms which burst out at the limit of the two atmospheres. The coast of the Pacific Ocean, from Punta Parina and Amatope to far beyond the tropic, from the equator to Chili, is scarcely ever refreshed by the rains of the ocean. Deprived of the vapors of the Atlantic by the chain of the Andes, these countries behold the vapors of the Pacific flitting away with the trade wind, and no accidental breeze brings them back. Drought and the desert are their portion, and on the border of the seas, in the very sight of the waves, they are reduced to envy the neighboring countries of the centre of the continent the gifts the ocean refuses to themselves, while lavishing them upon the others. Thus, under the same latitude, under the same tropical heavens, where the phenomena meantime are so regular, the two slopes of the Andes have a climate perfectly opposite. In one of them, the richest vegetation; in the other, drought, and a parched soil, the nakedness of which is poorly disguised by the light robe of a thinly scattered vegetation. The Andes separate the two climates by a sharply cut line, and testify strongly to the importance of the part performed in climates by the mountain chains, and their situation relatively to the general winds.

The northern and southern limits of this arid region are not where one would expect to find them at the first glance. The question is asked, why the same causes do not hinder the rains from watering the coasts of Peru, under the equator, and of New Granada. But, besides that the depression of the Cordilleras towards the north allows the trade wind to round it and to reach the western side, let us remember that this part of the coast corresponds nearly to the zone of calms, in which the direct influence of the trade wind is nearly annihilated, and where almost daily rain storms bring back to the earth the vapors in the very places whence they have risen. The influence of this latter circumstance here neutralizes the action of the Andes.

It is not the same at the southern limit. Here, not only the chain is continuous, but it forms, in the lofty table lands of Southern Peru and of Bolivia, the broadest and the highest terrace of all the Andes, shutting out all communication between the two sides. Moreover, we are here upon the limit of the tropic, and the regions in the neighborhood are often scantily supplied with rains, as we shall by-and-by understand. The lower regular trade wind begins, in fact, to blow there, and, as we know, the sky remains everywhere serene. The upper, or return trade wind, does not yet fall there; so that the causes of the condensation of vapor are wanting, and dryness of climate is the inevitable consequence. It is only at a greater distance, where the upper trade wind reaches the surface again, that the conflict of the winds commences, and with it the rains. On the coast of Bolivia, at the south of the Gulf of Arica, the two arefying influences unite and cause an almost absolute drought in the long desert of Atacama, which borders the coast nearly to Chili. It is only in the latter country, where the return trade wind of the north-west makes itself felt, that the rains recommence by degrees to water the earth.

In the part of South America situated beyond the tropical regions, the relative position of the Andes and of the plains on the east, produces an opposite effect. The vapors of the Pacific cannot penetrate there. The return, or north-west trade wind, avoids the coast and reënters the Atlantic Ocean, or, driven aside by the Andes, comes back arefied and made continental, across the plains of Paraguay and of the Pampas. Hence the violent west wind, in Buenos Ayres, called the *Pampero*, which carries to the coast only the whirlwind of dust it has raised in the arid plains it traverses in its course. The western coast, on the contrary, receives, with the return trade wind of the north-west, the vapors of the Pacific Ocean. Chili has rains in winter at the moment when the north-west reaches the neighboring regions of the tropics. More to the south, the winds of the sea coming from these parts, add their contingent, and give all this southern point of America the continuous rains belonging to the cool, temperate regions. Terra del Fuego and Cape Horn, at the confluence of all the sea winds, are incessantly bathed by the rains or covered by the snows; and the correctness of the not very flattering description Forster gives us of that climate, has been confirmed by all the navigators who have travelled through that inhospitable region of fogs and tempests.

Thus, in South America, the position of the plains and of the mountains, combined with the prevailing direction of the sea winds, produces the copious moisture of the tropical portion and the comparative dryness of the temperate.

In North America, an analogous disposition of the reliefs, and of the atmospherical currents, would doubtless produce the same dryness as in the plains of La Plata and the Pampas, if the deep cut of the Gulf of Mexico did not open the whole south of the continent to the wet winds of the tropics. Instead of coming from the interior of the continent, as in the temperate regions of South America, the return trade wind, which enters by this broad gate, comes directly from the seas, and has lost nothing of its vapors. It waters copiously along its course the whole Atlantic region and the western slope of the Alleghanies; even the valley of the Mississippi shares its benefits, although to a less degree. Towards the north, in the interior, the polar winds seem to resume their empire, and the moisture lessens. Thus North America is more favored with rains than could be expected from its situation westward of the return winds of the equator, and from its character as a large continent.

Along the western shore, from the coasts of Mexico to 60° of north latitude, we find the same succession of climates as in South America, in latitudes nearly corresponding. Between the tropics, in the rear of the high table lands of Mexico, where the trade wind of the Atlantic does not come, drought reigns, as on the coast of Peru. In the sub-tropical region, where the southwest trade wind has still but little influence, the rains are slight; they are almost none on the high table lands of California. Oregon, as well as Chili, has the winter rains, indicating the return of the upper trade wind to land; they seem to penetrate even beyond the Rocky Mountains, east of which the winter rains are frequent. Here we find the sources of the Missouri. In the North, finally, in Russian America, where the coast bends in and forms a deep bay, the south-west winds strike the coast, and produce the continuous and copious rains, the temperate, equal climate, and the vegetation of the coasts of Scotland and Norway.

The investigation we have just made of the distribution of the rain in the two Americas shows the influence of the direction of the high chains, and of their position on the western coast. It is immense. Place the Andes along the Atlantic, and the marine trade wind is arrested and dried; the table lands of Brazil, the endless plains of the Amazon, are nothing but a desert: no more of that wealth of vegetation, of those virgin forests, which now constitute their beauty; South America loses its character.

Place the Rocky Mountains east of North America, open the plains of the Mississippi to the south-west winds of the Pacific, and the climate becomes softer, more equal; the plains are still better watered, perhaps; nature has certainly changed. But what would then become of the present destinies, the entire future of this continent, were it necessary to cross the desert table lands of California, and their high mountain ranges, in order to reach the Mississippi from the Atlantic coast? What would become of its important relations with the Old World, if America, averted from the civilized nations, looked only towards the Pacific Ocean and China?

If we now direct our attention to the Old World, we shall again find the same influence of the forms of relief.

Tropical Africa, and the greatest part of the East of this continent, present two regions very unequally supplied with rains. On the north of the equator, the lands are less consolidated, the plateaus isolated from each other. Abyssinia is far from Mandara, and that is far from the Kong Mountains. The coast, from Cape Guardafui to that of Zanguebar, is slightly elevated; it permits the east winds of the Indian Ocean to penetrate the inland and to water all these parallels. The coasts of Senegambia and of Guinea are in the region of calms at the meeting of the two trade winds, and owe to this circumstance their copious rains, their climate, moist and fruitful, but treacherous and fatal to the man of the North.

On the south of the equator the plateaus are continuous; but instead of being in the West, as in America, the uplands are in the East; the eastern coast rises, and probably reaches, in the chain of Lupata, the loftiest elevation of this part of the continent. Then the eastern coast arrests the vapors; there the rains are everywhere abundant, from Cape Guardafui to Cape of Good Hope, while the vast elevated plains stretching from the west to the coasts of Congo, seem to exhibit, as far as we know them, only sterility and drought under the same latitude, where we see the plains of the Amazon and of Brazil drenched every year by torrents of water.

The contrast is complete; and whence comes this difference, if not from the disposition of the reliefs in the two continents?

The region of Cape of Good Hope is watered on the south-east coast, during the summer, by the winds of the Indian Ocean. But in the whole West the climate is dry except at certain points, and the Atlantic sends it only a few autumnal and winter rains.

The North, finally, Sahara, is closed towards the east against the access of the winds; its sub-tropical position and the nature of its soil contribute further to cause the deficiency of rain, making it one of the most vast and complete deserts in the world.

Western Europe, by its position, by the absence of high continuous chains along its seaboards, is open to the equatorial winds of the Atlantic, which bring their moisture thither all the year. The small extent of its surface, the number of its inland seas, and of the deep bays cutting into its mass, and leaving no place very far from some maritime basin; all these circumstances secure to it continued rains, mild climate, and that comparatively high temperature which belongs to it peculiarly. The numerous mountain chains, the endless diversity of soil, multiply the local condensations, as we have seen, and divide the continent into climatic regions as manifold as they are varied. Europe, alone, is without a desert.

In tropical Asia the monsoons and mountain ranges regulate the rains. The peninsula of India has the rainy seasons reversed on its two coasts; but its plentiful rains are reduced to a very small quantity on the plateaus of Deccan. All the region of Indo-China and of the great Asiatic archipelago is one of the best watered in the world. The conflict of the different winds, of which all this space is the theatre, the variety of the lands, so numerously scattered there, and the discontinuity of the chains, which can nowhere arrest the winds, are so many causes that secure to the whole of it such copiousness of tropical rains.

The Himalaya and the lofty chains of China stop the course of the ocean winds; all beyond, towards the interior, is a desert; it is the Gobi, the Tangout, and the sandy seas of Turkestan.

Australia is as yet so little known that it is impossible to analyze its climate. Nevertheless, what we have learned of late years concerning the configuration of its relief, proves that the highest lands, as in Africa, are placed on the eastern border of the continent. The trade wind of the Pacific scarcely penetrates thither, and that of the temperate regions shuns the coast. Furthermore, the southern half is, for the greater part, in the sub-tropical region, and seems to be deficient in mountains. Thus we may believe that the interior is a desert. But the eastern coast, Botany Bay, and the Australian Alps, are better watered than Swan River, on the western coast, and the prosperity of the colonies established on these two shores, has, of necessity, been in proportion. The mean quantity of rain water which falls in this part of the world is estimated at twenty-five inches; it is the most insular, and yet, owing to these circumstances, and to its rounded form, the most imperfectly watered of the continents. If what precedes did not inform us of this, the aspect and the slender forms of the vegetation, its attenuated leaves, which constitute its characteristic, would be sufficient to convince us of the fact.

Thus, gentlemen, if the *general climates* are given by the latitude, that is, by the spherical form of the earth, the *special climates*, characterized by the unequal distribution of the temperatures and the rains, are the effect of the grouping of the continents, and of the particular disposition of their reliefs.

In the point of view now occupying us, each continent has its special character. South America is the most humid of the tropical continents; North America, the best watered of the temperate continents, but the rains are equally distributed; Africa and Asia present the absolute contrast of dry and moist in the zone of the deserts touching upon the regions bathed by the rains of the tropics; temperate Asia is the dryest of the northern continents. Europe combines the moisture of the maritime climate with a great variety of contrasts; but they are all softened. Australia, finally, is the dryest and poorest of the continents.

The general law of the reliefs in the two worlds thus manifests its influence. The New World is that of plains, and the plains are open to the winds of the sea; its continental forms are less piled up and massive; it is, on the whole, the most humid. The Old World is that of plateaus and of vast extents; drought is its portion. It is enough to recall the influence these circumstances of humidity or aridity exercise on the vegetation, the aspect, and the organized beings of a country, to foresee that these great differences between one world and the other will be again reproduced in another province.

We have taken a rapid view of the variety of the phenomena to which the intermingling of the solid and liquid elements, of land and water, gives occasion. It would be easy, by a more detailed examination, to increase the number of these contrasts, of which I have pointed out only the most general. But I have said enough for a sketch of this vast subject, and to enable you to take a glance at all the wealth of life that nature unfolds by means so simple. I will add only one consideration more, which will serve for a conclusion to what we have thus far said of this great contrast of the continental and oceanic hemispheres.

We have seen, gentlemen, that it is from the combination of the two elements that life is born, a higher life than that belonging to either of them. It is neither the oceanic climate, nor the continental climate, which we shall proclaim as the foremost climate of the world; it is the combination of the two — it is the *maritime* climate. Here are allied the continental vigor and the oceanic softness, in a fortunate union, mutually tempering each other. Here the development is more intense, life more rich, more varied in all its forms. And when to these causes we further add the advantage of a tropical temperature, the forms of nature are, as it were, raised to their highest degree, and the wealth it brings to light surpasses all elsewhere seen. I will cite only a single example: this will suffice.

Nowhere on the surface of the globe is the blending of the continental and oceanic element so complete, and on so great a scale, as in the East Indies, and in that archipelago—the greatest in the world—which fills the space comprised between the South of Asia and Australia. Peninsulas, which are worlds, as those of Deccan and Indo-China; islands, which are small continents, like Borneo and Sumatra; a blending of chains and plateaus, and of plains, as on the continent; and all this cut up, bordered, or surrounded by seas in the most diversified manner, bathed by the humid atmosphere of the tropics, and exposed to the burning rays of a vertical sun—*these* are all the means of physical life which nature can receive. And then, what mighty, what admirable vegetation! We see at the same time plants with broad and numerous leaves, the excessive expansion of which is always the proof of an exuberant humidity; and those shrubs with concentrated and elaborated gums, those spices, those aromata, that bear witness to the dry and intense heat of the continent. There is the country of the mighty Banian, the symbol of vegetable strength. There uplifts its head the majestic Talipot palm, a single leaf whereof, sixteen feet broad and forty feet round, is enough to give shade to a score of men at once; and in the bosom of those virgin forests grow the largest flowers in the world, the Rafflesia, whose gigantic corolla alone measures no less than three feet across. There grow the cinnamon, the nutmeg, the pepper, and the cloves, which all civilized nations have fetched thence from time immemorial.

Everything most grand and powerful of the productions of the animal world is there encountered. The rhinoceros, the huge royal tiger, the orang-outang, that great serious-looking ape, the most perfect of animals, and that which seems to foreshadow in its structure the complete configuration of the human body, are all inhabitants of those countries. If to these we add the mineral wealth, the gold and the diamonds, abounding there, we may pronounce these regions the most richly endowed of the universe.

But let us raise ourselves above the limits of the natural, into the regions of the historic world. Where have we beheld all peoples and societies arrive at their highest perfection, if not in Europe, that peninsular continent, the most indented and most maritime of all the continents? Where do we see barbarism reign triumphant, if not in Africa and Australia, continents shut off from all contact with the rest of the world, its seas and its people, by their continuous and unindented outlines? This is neither the time nor the place to analyze the causes of this phenomenon; I now merely allude to the facts, intending to return to the subject hereafter. But I will add, that it is not an isolated fact. Call together your historical recollections, and cast your eyes upon this map of the world, and you will see that all the highly civilized peoples of the earth, with the exception of one or two primitive nations, have lived, or still live, on the margins of seas or oceans.

The Chinese and the Hindoos unquestionably represent the most advanced state of civilization in Oriental Asia. In Europe, to name Phœnicia, Asia Minor, Greece and Rome, is to enumerate all the highly cultivated peoples of antiquity, and all have, as the theatre of their strifes and exploits, as well as for their connecting link, the Mediterranean Sea. To come to a later date, it is to the ocean that Spain and Portugal owe the brilliant part they played, at the period when superb discoveries doubled the extent of the historic world. At this very hour, to conclude, the might of England causes itself to be felt from one to the other extremity of the world.

And in this new world of North America, now entering on its great career among the nations under so happy auspices, is it not on the shores of the Atlantic that life is developed in its most active, most intense, and most exalted form? Is this merely a chance consequence of the accidental debarkation at that point of the colonists of the Ancient World? No, gentlemen, brilliant as may be the prospects the West may aspire to from the exuberance of its soil, life and action will always point toward the coast, which can only derive fresh accessions of prosperity from the prosperity of the interior. The life of nations is in the commerce of the world, not only in a material, but even more in a moral point of view; and it is because America is enthroned queen-like upon the two great oceans, that she will be called to play a part as mediator between the two extremities of the world, of which no one can at this moment conceive the magnificent extent.

This, then, is the resolution of the contrast between

the continental and the oceanic world, as regards the intermixture of their natures. It is in this region of contact between the sea and the ocean that life is unfolded in its most intense and diversified form; and, both in point of nature and of history, the maritime zone of every continent enjoys a superiority over all others not to be questioned or disputed.

LECTURE VIII.

The marine currents — The motion of the seas due to other causes than that of the continental waters — Various causes of the marine currents — Differences of temperature the principal, acting indirectly by the winds, directly by the unequal density of the waters — Coincidence between the great atmospheric currents and the marine currents — System of general currents — The Equatorial current and the Polar currents — The currents of the Pacific Ocean; of the Indian Ocean; of the Atlantic Ocean — Contrast of the Old World and the New — Disposition of their continental masses — Consequences — The Old World the continental; the New the oceanic — The first essentially temperate, the second tropical — Special character of the New World — Its structure more simple — Abundance of its waters — Vegetation predominates on the Animal World — Incomplete development of the higher animals — Influence on the indigenous man — Conclusions.

LADIES AND GENTLEMEN: —

Thus far we have been studying, in the grand contrast of land and water, the influence of the oceanic element on the continental world, of the wet upon the dry, more than that of the continents upon the oceans. This was the right order, because the continental world is much the more important. Nevertheless, we will not leave the subject without saying, at least, a few words upon the action the continents, in their turn, exercise upon the oceans. Now, as this action of the lands is essentially limited to regulating and modifying the movements of the oceanic waters, by their disposition, by the forms of their coasts, and their submarine relief, it will be sufficient for our purpose to take cognizance of the principal phenomena presented by the marine currents, without entering upon details which the shortness of the time allowed compels me to pass over in silence.

The spectacle around us has accustomed us to see the continental waters in motion. We hear, without astonishment, the streams murmur in the meadows, the torrents roar in the mountains, and behold the rivers flow majestically along their bed. The cause of this motion of the water is familiar. We know that the particles of this movable element, influenced by the hidden power of gravitation, move and flow untiringly, until they have reached the lowest place accessible to them. If in their course they fall in with a basin having no exit, they gather there and put themselves into a state of equilibrium and repose, preserving their horizontal surface and their immobility, until driven from it by another force foreign to the first. Such are our peaceful lakes, with pure and tranquil waters, whose mirror reflects the mountains that adorn their margin, the azure of the sky, and even the slightest cloud floating in the atmosphere that bends over them.

But these basins, which here and there collect the living waters of the continents, are only the image, on a very small scale, of what the vast and deep basin of the oceans is for the whole of the waters of our planet.

All the water springs that furrow the continents tend towards this common reservoir. Gushing from the height of the table lands, or the lofty summits of the mountains, they pour their waters first from fall to fall headlong down the rapid slopes; they traverse, at a more sober and measured pace, the long low plains leading to the ocean, into which, as we behold their slower and slower and more sluggish march, their waves seem unwillingly to enter, as if conscious that they were to be confounded together, and lose their existence there. Here, in truth, ends the ephemeral life of the rivers; their motion has ceased, they disappear in the immensity of that vast abyss whence they had issued.

We are, then, inclined to look upon the basin of the oceans as containing waters in a state of rest; for the cause which sets the river waters in motion exists no longer; the differences of level are annulled. Yet let us be cautious; all this may well be only a first appearance. The very mobility of water, which prevents it from reflecting permanent forms, and levels all inequalities, renders it also accessible to the slightest external influences, and several causes succeed in impressing upon this element, passive in the highest degree, the most varied motions.

The winds raise the waves of the ocean by an action wholly mechanical, and produce only a superficial and local agitation; but when they blow constantly in the same direction, they impart to the waters a transfer motion in the direction of their own course. The sun and moon pass over the surface of the seas, and the entire mass of waters, obedient to the mighty attraction, piles itself up in a vast swell, whose summit follows the course of the dominant luminary. These are the tides. The unequal pressure of the atmosphere on the different points of the ocean, whence result differences of level, and, above all, the differences of temperature between the tropical and the polar seas, to which correspond different degrees of density, are so many more causes disturbing the equilibrium of the oceanic waters, and creating in their bosom various motions which continually tend to reëstablish the equilibrium, but without effecting it. Sometimes the superficial mass is transported from east to west, as in the great equatorial current; sometimes a deep and narrow band, a true oceanic river, flowing rapidly through waters comparatively tranquil, as the Gulf Stream. Here the currents meet and unite; there they are superposed, and the upper and under currents run in opposite directions. Everywhere is agitation; nowhere absolute rest, as unknown to nature here as in all other quarters.

The greater part of the causes, to say nothing of others more doubtful which it would be useless to mention here, often act in concert to produce marine currents; but it would be difficult to assign to each of them the exact portion of effect belonging to it. There is one, however, which seems to control all the rest by its power and the constancy of its action, direct or indirect, and that is the difference of temperature between the regions near the equator and those in the neighborhood of the poles. Now, since the general winds, as we have seen, owe their origin to this same cause, we shall not be surprised to find a similarity, and, in some cases, a remarkable coincidence, between the march of the great atmospheric currents and that of the general currents of the ocean. For, not only do the winds act directly on the currents and sweep them forward in their course, but the same forces drive them both in a common direction; the same obstacles, the continents, check their onward movement, and force them to swerve, in a similar manner, from their original direction. A knowledge of the one will facilitate the understanding of the others.

The most general fact to be noted here is the existence of the great equatorial current, as it has been agreed to call it, which seems a general transfer movement of the tropical waters from east to west all round the globe, rather than a current properly so called. This grand phenomenon did not escape the sagacity of Columbus, who was also the first to discover it. "It seems beyond a doubt," said he, after one of his earliest voyages, "that the waters of the ocean move with the heavens;" that is, in the direction of the apparent course of the sun and stars. This great current is analogous to the trade winds; it has ever been thought that these winds were the principal cause of it. But it is too deep and rapid to admit of being explained by their action alone.

The difference of temperature between the tropical and polar seas, and the loss the seas of the warm regions suffer from more active evaporation, would be a still more profound and irresistible cause. The colder and heavier waters of the polar regions perpetually tend to flow towards the warm and lighter waters of the equator, and to displace them. The existence of these polar currents is demonstrated by the floating masses of ice which, swept on by the waters whence they had their being, accomplish every spring long pilgrimages towards the warmer regions, and stray even as far as the 40° of latitude. Like the atmospheric currents setting from these same quarters, they occupy the lower part of the domain of the oceans, while the warm waters of the equator spread over their surface. Hence the astonishing spectacle of those majestic ice-bergs, of which only an eighth part is visible, while the rest is sunk in the depths of the sea, continuing their solemn progress southward, and, on meeting the Gulf Stream, moving on in a direction opposite to the course of its waters, proving thus that the waters enveloping their bases pursue without obstruction their southward course. The polar currents, while advancing towards the equatorial regions, gradually make a bend west-ward, like the winds, under the influence of the earth's rotation, and at the meeting in the tropics are trans-formed into a vast movement from east to west. Add to this general tendency of the deep waters, the direct and constant action of the trade winds upon their surface, and that of the tides acting in the same direction, and the cause of this phenomenon will appear to you, if not fully explained, at least sufficiently accounted for.

The grand equatorial current is still more disturbed than the trade winds, by the continents arresting their progress and causing the waters to flow back in very different and often opposite directions to their origi-nal course. Each of the three great oceans forming a separate basin, and presenting a collective combination of physical circumstances which modify the march of the oceanic currents in a peculiar manner, we proceed to pass them in review successively, beginning with the Pacific Ocean, whose system is more simple than that of the two others. (See plate I.)

The Pacific Ocean, owing to its vast extent, gives full and unimpeded sweep to the general currents in a more regular manner than either of the others. The Antarctic polar current, bent eastward by the prevailing winds of these regions, strikes the western coast of America between 50° and 40° south latitude. It divides into two branches, of which one runs southward, doubles Cape Horn, and carries its waters on to the Atlantic. The second and principal passes along the coast of Chili and Peru, cooling the climate by the low temperature of the waters it bears, which are from 10° to 12° Centigrade, or from 18° to 22° Fahrenheit colder than the neighboring sea off Lima. The current, to which it has been proposed, on good grounds, to give the name of Humboldt,—who was the first to prove its origin and abnormal temperature,—suddenly quits the coast near the height of Punta Parina, and goes on to form the grand equatorial current.

This latter occupies a breadth of nearly 50 degrees on the two sides of the equator, and passes beyond the tropics, north and south. It follows its peaceful and majestic course, unobstructed, with an average speed of from 30 to 35 miles a day, to the chain of islands that fringe the continents of Asia and Australia. On the north it reaches Formosa, and, running upon the coast of China, turns off and moves to the north-east along the shores of Japan. On the south, it is already disturbed by the monsoons, and loses its way in the labyrinthine mazes of the grand Asiatic archipelago, whose seas heave with the violent currents that add further to the dangers of navigation in these stormy seas.

In the northern part of this ocean, the west winds reigning there determine a drift current, which advances to the American coasts and conducts the waters southward along the shores of California, whence they doubtless reënter the equatorial current, to commence a new circuit.

The polar currents seem to be almost nothing. The bank or neck, which in all probability unites the neighboring points of the continents of Asia and America under the waters of Behring's Strait, hinders the under currents coming from the pole from entering this basin, while the warmer waters of the Pacific flow on the surface into the Frozen Ocean.

In the Indian Ocean, the equatorial current, like the trade wind, is broken. In the region of the monsoons, or the northern region, the currents follow alternately the direction of these periodical winds, and flow with them six months in one direction, and six months in another. But in the south, where the trade wind retains its empire, the normal current holds its way, narrows as it approaches Madagascar, sweeps north of that island, and, stemmed by the coast of Africa, enters the channel of Mozambique.

Compressed into this narrow passage, it acquires the enormous speed of four or five miles an hour, and, reinforced south of Madagascar by another branch, reaches in its rapid course the Cape of Good Hope, and the great Needle bank, las Lagullas, the borders of which it follows at a distance from the coast. Here it divides. One part encounters the current setting from the south ern Atlantic, and with it reënters the Indian Ocean The other branch doubles the Cape, enters the Atlantic, and, flowing along the western coast of Africa, proceeds to blend its waters with those of the equatorial current of this third ocean.

The forms of the Atlantic Ocean, so characteristic, the small breadth it presents in the region of the equator, the deep windings of the Caribbean Sea and the Gulf of Mexico, wherein nearly all the tropical waters of this ocean are accumulated, as in a receptacle having no outlet, give to its currents an aspect both more marked and less normal. The equatorial current does not there assume its customary proportions, while the return current, the Gulf Stream, is exhibited in a very unusual manner. These are the two salient features necessary to study first.

The equatorial current connects itself with the current of the Cape of Good Hope issuing from the Indian Ocean. Starting from the coasts of Southern Africa, it soon extends both sides of the line, widens considerab y, and flows across the ocean at the rate of two to three miles an hour. Having reached the coasts of America at Cape Rocca, it divides, one branch flowing southward, along the coasts of Brazil, and, together with the waters of the southern basin, resumes the route of the cape and the Indian Ocean. The other and principal branch takes a west-north-west direction, rolls its waters along the shores of Guyana, enters the Caribbean Sea, which Rennel calls a sea in motion, rather than a current, penetrates into the Gulf of Mexico, making its circuit, and, passing before the mouths of the Mississippi, arrives at the narrow passage between the point of Florida and the Island of Cuba, whence it comes forth under another name.

In truth, the accumulated and moving waters of the Caribbean Sea and the Gulf of Mexico are the inexhaustible source of that torrent of warm water, which, under the name of the Gulf Stream, precipitates itself over the breakers of Bahama, flows along the coast of Florida, at a rate varying from two to five miles an hour, according to the season, and keeps on its way, upon a line parallel to the shore at a short distance from its margin, until it passes beyond Cape Hatteras. The Stream, hitherto narrow, deep, and rapid, meets in this vicinity the cold waters from the north, and the sand banks running along at a distance from the coasts as far as the southern part of Newfoundland Repulsed by these obstacles, it makes a sudden turn to the east, becomes much broader, spreads over the surface, and holds henceforth its slackened course to the Azores, whence it bends towards the south, in order to recommence from the coasts of Africa the immense cycle of its never-ending rotation.

These warm waters of the tropics advance northward even beyond the limits we have just indicated. Driven by the south-east winds prevailing in the northern Atlantic, they proceed to bathe the coasts of the North of Europe, the temperature of which they soften, and often deposit on the lonely shores of Scotland and Norway the plants and seeds of the tropical regions, — unanswerable witnesses of their distant course.

On seeing the narrow breadth of the Gulf Stream from its origin to Cape Hatteras, one is led to ask how it can be sufficient to cover with warm water the immense surface it occupies from this point all the way to the Azores. The beautiful explorations, executed under the able direction of Professor Bache, Superintendent of the Coast Survey, give the answer; for numerous thermometrical soundings prove that off this cape the depth of the current is such, that at 3,000 feet below the surface it still presents nearly the same differences of temperature which distinguish it from the surrounding sea, and clearly mark its limits. It is doubtless these deep waters that appear at the surface when it becomes broad; for as it loses in speed, the warm waters are free to ascend and take the place assigned to them by their lesser density, at the same time this very cause favors the accumulation of the waters in the part of the current where its progress is slackened. It only changes form, and, in advancing, must lose in depth what it gains in width.

The polar currents of the Atlantic are perceptible chiefly on the coasts of America. Hudson's and Baffin's Bay and the Sea of Greenland pour their waters and their ice along the eastern coast of the continent, and contribute, doubtless, to lower the temperature.

Such are the most salient features of the vast picture presented by the oscillations of the ocean waters. Although we have merely touched upon the subject, we know already enough, I believe, to be convinced that, if the causes of these movements flow, for the most part, from the general laws regulating the physical constitution of the globe, their evolutions, and the special and individual characters they assume in each ocean, are an immediate result of the configuration and disposition of the terrestrial masses forming the basin of the seas.

The great oceanic currents are one of the grandest phenomena presented by the wise economy of nature. Their extent, the prodigious length of their course, in some nearly equal to the circumference of the globe, fill us with astonishment, and leave far behind everything of this description to be seen in the water courses of the continents. Owing to these permanent streams, the sea waters mingle from pole to pole, and move with sleepless flow from the Pacific to the Indian Ocean, and from this to the Atlantic; and this unending agitation preserves their healthfulness and purity.

Like the winds, the currents tend to equalize differences, to soften extremes. The cold waters of the Antarctic pole temper the scorching heats of the coast of Peru; the warm waters of the Gulf Stream lessen the severity of the climate of Norway and the British islands. Their importance is no less in the relations of the people and the commerce of the nations. It is the currents which, together with the winds, trace the great lines of communication upon the highways of the oceans, favoring or obstructing the intercourse of one country with another, bringing near together places apparently the most remote, separating others that seem to touch each other. Their importance in nature and history cannot fail to impress the minds of all.

We abandon the ocean, and shall, henceforth, occupy ourselves only with the continental masses. To study them better in their analogies and their differences, to detect their true character, we shall consider them successively in their natural groups, under two different points of view, which we have already indicated; I mean as the Old World and the New, then as the Northern and Southern continents. Let us begin with the contrast of the Old World and the New.

The most prominent feature of the arrangement of the continents is, in fact, the grouping of the two Americas in one hemisphere, and that of the four others in another hemisphere. This division of the continents into two worlds is so evident from the first glance, and is at the same time so convenient in practice, that it has passed into common speech as one of those observations admitting no contradiction.

But to bring out prominently the contrast of these two worlds, they must be studied more in detail than we have thus far done; we must compare them, in order to deduce, by the comparison itself, the special character of each. This is what we are going to attempt. We have already seen that they differ in the forms of their relief and in their climate; we shall further see that these fundamental differences produce analogous effects in the organized beings, and in the entire physical life of each of the two worlds. Finally, we shall speak of the manner in which they act upon each other, and seem, by their very nature, destined not to live isolated, but to form together a single organism, a grand harmony.

During the whole of this study, please to remember, gentlemen, we are in the realm of nature, and not in that of history. The America we are seeking to understand is that which Columbus and his successors discovered, still entirely a virgin world, centuries ago; and not the New World of history, we shall have to speak of later, that has come to plant itself on that soil.

A general comparison of the two groups of continents will call to mind some of the leading features we have already become acquainted with, and add some others.

The Old World and the New World differ in the groupings, and in the number and extent of the continents composing them; in their astronomical situation, with respect to the climatic zones; in the general direction of their lands; in their interior structure. This assemblage of opposite characters secures to each of them a climate, a vegetation, and an animal kingdom, peculiarly their own.

I say, first, in their groupings: —

The Old World is composed of four continents. Setting aside Australia, which is only an island in the midst of the oceanic hemisphere, it numbers three, all very near each other, aggregated, and forming an oval, compact mass, whose extent far surpasses that of every other terrestrial space. It presents a solid extent of land, the most vast, the most unbroken, the least accessible in its centre to the influences of the Ocean. The Old World is preëminently the *continental* world.

The New World has only two continents, North America and South America, America and Columbia, as I should like to call them — to render justice where right belongs—if it were not forbidden to change names consecrated by long usage. These two continents are not grouped in one mass, nor placed side by side, but separated from each other, not touching upon their long sides, but by their exterior angles, standing in line, rather than grouped. They are situated in two opposite hemispheres, and thus more distant from each other, apparently, and less neighboring.

The result of this remarkable disposition is that narrow, lengthened, slender form we see in the New World. No portion of the interior is very remote from the sea-coasts; everywhere it gives access to the influences of the ocean, in the midst of which it is placed, like a long island. This form already, contrasted with that of the Old World, gives to it its character. The New World is essentially *oceanic.*

The astronomical position, relatively to the climatic zone, is also not the same in the two worlds.

The Old World is, as it were, crowded back upon the north of the equator; it belongs, for the most part, to the northern hemisphere and to the temperate zone. Of the three principal continents composing it, the two whose importance is by far the greatest, Europe and Asia, are temperate. Asia penetrates the torrid regions only by the southern peninsulas; Europe at no point; Australia is sub-tropical; Africa only is truly tropical. Even if we take in the last two continents, more than two thirds of the lands are situated in the temperate regions, one third only in the equatorial regions. The Old World is then essentially *temperate.*

In the New World the lands are distributed in a manner nearly equal in the two zones and in the two hemispheres. We find that the countries it includes, those the most richly endowed, are situated under the sun of the tropics. Compared with the Old World, the New World is thus essentially *tropical.*

The general direction of the lands, or the direction in which their length extends, is the inverse in the two worlds. The Old World has its greatest prolongation from east to west, in the line of the parallels; the New World from north to south, in the direction of the meridians. Both have a length of about 7,500 miles, but the breadth of the Old World is nearly double that of the New. This disposition is of the greater consequence for the distribution of the climates in each of them, since this configuration coincides, as we have seen, with the interior structure, with the direction of the principal mountain chains, and of the table lands. From one end to the other of the Old World, over a space of several thousand miles, the migratory tribes are able to pursue their adventurous roaming course by following, according to their custom, the great features of relief of the soil, without witnessing any change in the vegetation or the animals that surround them. They change place but not climate, nor ways of life. This similarity of climates over long spaces is, then, a property of the Old World, and must have singularly favored the dispersion of the primitive tribes.

In the New World, on the contrary, the zones of similar climates are short and numerous; and if we travel over the whole length of the two Americas, we pass twice in succession through all the temperatures, from the frozen climate of the poles to that of the equator, and from the burning climate of the equator to that of the poles. This diversity of climate gives their character to the Americas.

Meantime, the interior structure modifies these climatic relations in the two worlds, in such a manner as to correct the uniformity of climate in the Old by more marked contrasts, and the too great diversity of the New, by gentler and better graduated transitions. We shall see this as we proceed to a closer examination of the structure of the two Americas, which will particularly occupy us this evening.

The maps we shall make use of in this inquiry, these you see before you, require perhaps some explanation. They are intended to enable the eye to take in by a single glance the different elevations of relief; we see here the low plains, the table lands, and the mountains, each indicated by a particular color. (See plate I.) I need not call your attention to the usefulness of substituting, in teaching geography, such physical maps, for the flat and perfectly unmeaning charts found in the common atlas.

What characterizes the interior structure of the New World is its simplicity. In place of the variety of the Old World, where, in spite of a few general features common to both, each continent is, as it were, cast in a separate mould, the two Americas seem absolutely formed upon the same plan. This plan may be sketched out in a few lines. Two triangles, their vertices turned to the south, one situated north-west of the other; the long cord of the Rocky Mountains and of the Andes, running the length of the extreme western coast, and binding the two continents together; great plains on the east, forming the larger part of their surface; a slightly elevated chain along the Atlantic coast of both, the Alleghanies in North America, the Serra do Espinhaço and the Serra do Mar of Brazil, in South America; finally, in the centre, three short, transverse chains, that of Parime in the Guyanas, that of Venezuela, and that of the great Antilles, broken into a number of islands;—these, in a few words, are all the essential features of this vast division of the world.

That which constitutes the richness of organization in the continents, is the number and abundance of internal contrasts calling out at once the activity of nature and that of man. The Old World is full of them; America has only a small number, all tending to disappear by reason of the structure itself.

Thus, in Asia and Europe, the line of the highest lands, the continental axis, extending from the Himalaya to the Alps and the Pyrenees, divides these two continents into two unequal parts, one north and one south, opposite in climate, in vegetation, and even in races. Scarcely anywhere is the transition from one to the other gradual; almost everywhere it is abrupt and sudden. The table lands of Tubet and frigid Mongolia touch the tropical plains of China and of the Indies; the traveller who passes the Alps, abandons the severe landscapes and the firs of the North, to descend, by a single day's journey, into the ever verdant gardens and the orange groves of fair Italy; he exchanges the cold mists of the North for the sun of the South, and often eaves on one side the snows and frosts of winter, to find on the other the warm breath of spring, its verdure and its flowers.

This striking contrast between the North and the South, reflected in the character and history of all the nations of Asia and Europe, is doubtless found in America; it is perhaps too well known in this country. But in nature it is almost effaced; is softened down. It does not form a barrier; nowhere presents an abrupt change; nowhere breaks the unity. On account of the continued plains of the continent, we see the natural character of the North gradually melt into that of the South. Between the shores of the Frozen Ocean and the Gulf of Mexico lies the whole distance from the icy regions to the tropics. But it is only slowly, and over long spaces, that we pass through the transition. From the treeless polar plains, where flows the Mackenzie River, whose sole covering is the mosses and the lichens, we pass only by degrees to the coniferous forests of Lake Superior; then to the oaklands of Wisconsin; the walnuts, hickories, and the chestnuts of Ohio and Kentucky next appear; further south, the magnolia and the palmetto already herald the air of the tropics and the neighborhood of the Gulf of Mexico. Two thousand four hundred miles separate the extremes of this scale of vegetation, which almost touch each other in the Himalaya.

It is, moreover, to these vast plains, offering no obstacles to the dissemination of the species, and to the absence of great chains from east to west, that we undoubtedly owe the appearance, at the North, of plants and animals that seem to belong only to the tropical regions. Not without surprise, the European, landing on these shores, sees the humming-bird, that diamond of the tropics, glancing in the sun in a country which winter clothes, during long months, in a thick mantle of snow and ice.

It is the same towards the South, where we see the palm trees and the parrots of the tropics, here and there, as far as the Pampas of Buenos Ayres, much beyond their natural limits.

Thus the contrast between the North and the South is softened, reduced; but it is not annihilated; it exists on a great scale from one of the continents to the other; for North America is temperate, and South America is tropical.

America is cut by the Andes into two parts, East and West, as Asia and Europe are cut into two parts, North and South. But this contrast likewise is almost neutralized, as we shall soon understand.

The inequality is here carried to the extreme, to such a reduction of one of the parts, that it loses its importance, and, so to speak, its power of reaction. The western coast, dry and barren, has not extent and influence enough to enter upon an effective rivalry with the vast countries of the East. Moreover, the difficulty of communication renders the mutual action and the intercourse between the countries situated at the foot of the two inclinations, still more rare. Finally, the two sides of the Andes, being under the same latitudes, have the same, or nearly the same, climate, and differ merely in the degree of moisture or dryness falling to their portion. The West of the two Americas is only a narrow strip, not to be compared with the great plains of the East occupying nearly the entire continent, and giving it its character.

America is then less rich in internal contrasts than the Old World, but has more of unity, because it is more simple. Undoubtedly, in this uniformity of structure, in this absence of obstacles to a free circulation from end to end of this world, we are to look for one of the principal causes of that common character, of that American physiognomy, which strikes us in all the organized beings of this continent, and which we find again in man himself, in the Indian, all the tribes of whom, from the banks of the Mackenzie River to Patagonia, have the same coppery tint and a family likeness in the features, impossible to mistake.

The climate of the New World, compared with that of the Old, is distinguished by the abundance of pluvial waters, in general, by a greater humidity. We have seen in what manner this phenomenon is the consequence of its narrow and lengthened form; of the opening of the great plains—that is, of the two continents almost entire—to the winds of the sea; of the absence of high mountains in the East; in a word, of the configuration and general exposure of this part of the globe. While the Old World, with its compact figure, its vast plateaus, its high lands in the East, receives only an average of 77 inches of water by the year under the tropics, America receives 115 inches.

The temperate regions of Europe have 34 inches; North America, 39 inches.

Add to this abundance of water, the extent of plains permitting the development of vast systems of water courses, and you will understand the existence of that innumerable multitude of rivers and lakes, which are one of the most characteristic features of the two Americas. Notwithstanding a much smaller extent than that of the Old World, the New possesses the largest rivers on the earth; the richest in waters, those whose basins occupy the broadest areas. Where can we find, on the surface of the globe, a river equal to the mighty Marañon, that giant among the rivers of the earth, gathering its waters from a surface of a million and a half square miles, and bearing them to the ocean, after a course of 3,000 miles? This mighty monarch receives in his progress the homage of tributaries, each of which, by its greatness and the abundance of its waters, would suffice for the wants of a whole vast country. Such are the Ucayale, the Rio Purus, the Rio Negro, above all the Madera, rivalling in importance the river to which it yields the honor of giving a name to their united waters. The further it advances in its majestic course, the more its proportions increase; and before arriving at the ocean, its broad sheet, from the middle of which the eye cannot reach the banks, seems rather to be a fresh-water sea, flowing sluggishly towards the ocean basin, than a river of the continent. Far from its mouth, the fresh and muddy waters of the Amazon are still distinguished at a glance of the eye from the saline and limpid waves of the ocean; and the slime, borne along by their currents, goes to form, further north, a new soil on the shores of the continent.

In the other continent, the Father of Waters, the mighty Mississippi, the second of the rivers of the earth, equals in length the Marañon itself; for its winding course is 3,000 miles. But its basin covers only from 8 to 900,000 square miles. Who does not know the importance of tributaries like the Missouri, which wrongly gives up its name for that of its less powerful brother; like the Ohio, the Beautiful River, the stream with transparent waters; like the Arkansas, and so many others composing that vast system of arteries that vivify the whole West, and are destined to assume daily a greater and greater importance? And these immense rivers are not isolated. At the side of the Marañon, the La Plata has a course of not less than 1,900 miles, and more than a million of square miles send into it their waters. At the side of the Mississippi, the St. Lawrence has a course of 1,800 miles, and a basin of nearly a million of square miles.

The Old World offers nothing similar. The greatest of its rivers, the Yan-tse-Kiang in China, has a course of only 2,500 miles. The Ganges and the Nile are far from equalling it. The Volga, the greatest of the rivers of Europe, exhibits a course of only 1,700 miles, and if it were necessary to enumerate in America rivers like the Rhine, so celebrated, it would be almost by the hundred that we should have to cite them.

And what shall we say of the abundance of its lakes? The group of the great lakes of Canada, so characteristic of North America, nowhere finds a parallel. It contains at once the largest lakes in the world, and the greatest mass of fresh water united on the surface of the continents. These vast fresh water seas, together with the St. Lawrence, cover a surface of nearly 100,000 square miles, and it has been calculated that they contain almost one half of all the fresh waters on the surface of our planet. They, too, are not alone, and a glance of the eye at the map enables us to perceive in the North a multitude of lakes but little inferior in extent: the lake Athapescow, Winnipeg, Slave Lake, the Great Bear, are worthy to figure side by side with the lakes of Canada and of the St. Lawrence.

The rivers and the lakes are the wealth, and justly form the boast, of America. No continent possesses so great a number, or those of such large extent, so well supplied with water, so navigable. Not only do they fertilize the rich countries they traverse, but they are now, and will become still more so, the great highways of commerce between all the parts of this vast world; we already see enough to hope everything of the future.

Thus, gentlemen, the watery element reigns in the New World; add to this, that half of its lands are exposed to the rays of the tropical sun, but that, all the conditions being equal, America is, on the whole, a little less warm than the Old World, and we shall have the essential features of its climate. The oceanic climate—

this is what America owes to the fundamental forms and the relative disposition of its lands; while the Old World is indebted to it for the preponderance of the dry and continental climate.

Let us now follow out the consequences of these physical circumstances upon the development of organic beings, and the character of the New World will come out in all its clearness.

The warm and the moist—these are the most favorable conditions for the production of an exuberant vegetation. Now, the vegetable covering is nowhere so general, the vegetation so predominant, as in the two Americas. Behold, under the same parallel, where Africa presents only parched table lands, those boundless virgin forests of the basin of the Amazon, those selvas, almost unbroken, over a length of more than 1,500 miles, forming the most gigantic wilderness of this kind that exists in any continent. And what vigor, what luxuriance of vegetation! The palm trees, with their slender forms, calling to mind that of America itself, boldly uplift their heads 150 or 200 feet above the ground, and domineer over all the other trees of these wilds, by their height, by their number, and by the majesty of their foliage. Innumerable shrubs and trees of smaller height fill up the space that separates their trunks; climbing plants woody-stemmed twining lianos, infinitely varied, surround them both with their flexible branches, display their own flowers upon the foliage, and combine them in a solid mass of vegetation, impenetrable to man, which the axe alone can break through with success. On the bosom of their peaceful waters swims the Victoria, the elegant rival of the Rafflesia, that odorous and gigantic water lily, whose white and rosy corolla, fifteen inches in diameter, rises with dazzling brilliancy from the midst of a train of immense leaves, softly spread upon the waves, a single one covering a space of six feet in width. The rivers rolling their tranquil waters under verdurous domes, in the bosom of these boundless wilds, are the only paths nature has opened to the scattered inhabitants of these rich solitudes. Elsewhere, in Mexico and Yucatan, an invading vegetation permits not even the works of man to endure; and the monuments of a civilization comparatively ancient, which the antiquary goes to investigate with care, are soon changed into a mountain of verdure, or demolished, stone after stone, by the plants piercing into their chinks, vigorously pushing aside, and breaking with irresistible force, all the obstacles that oppose their rapid growth.

South America, and particularly the basin of the Amazon, is the true kingdom of the palm trees; nowhere does this noble form of vegetation show itself under a greater number of species. This is a sign of the preponderating development of leaves over every other part of the vegetable growth; of that expansion of foliage, of that *leafiness*, peculiar to warm and moist climates. America has no plants with slender, shrunken leaves, like those of Africa and New Holland. The Ericas, or heather, so common, so varied, so characteristic of the flora of the Cape of Good Hope, is a form unknown to the New World. There is nothing resembling those Metrosideri of Africa, those dry Myrtles (Eucalyptus) and willow-leaved acacias, whose flowers shine with the liveliest colors, but their narrow foliage, turned edgewise to the vertical sun, casts no shadow. Everywhere, long and abundant leaves, an intense verdure, a strong and well-nurtured vegetation, these are what we find in tropical America.

North America, in spite of its more continental climate, shares no less in this character of the New World. The beauty and extent of the vast forests that cover its soil, the variety of the arborescent species composing them, the strong and lofty stature of the trees growing there, all these are too well known for me to stop to describe them. It is because this continent adds to a more abundant irrigation a soil slightly mountainous, almost everywhere fertile, securing to it always an equal moisture, a more abundant harvest of all the vegetables useful to man.

Not only is the vegetation abundant in the New World, but it is universal, and this is a further characteristic distinguishing it from the Old. We do not see there those vast deserts, so common in the other continents, and occupying a considerable portion of their surface. The deserts of California and that of Atacama are exceptions, and, compared with those of Africa and Asia, scarcely seem made for anything except to serve as specimens. The llanos of Orinoco, which their geological nature dooms apparently to the fate of Sahara, are copiously watered during the rainy season, and are covered then with an admirable vegetation. Life, which seemed almost to slumber, almost extinguished, springs up again more beautiful and more vigorous.

To the powdered sand, swept along by the winds, succeed rich pastures, where range a multitude of indigenous animals, mingled with herds of horses, and wild asses, coming from Europe; and thousands of reptiles, buried in the watery slime during the dry season, reappear, and fill again with life the temporary rivers and lakes with which these valleys are then overflowing. The pampas themselves are not without vegetation, and support at all times numerous herds. And who is ignorant that the endless prairies of the Mississippi and the Missouri produce every year an abundant vegetation, on which feed the bisons and the other wild tenants of the countries?

But what becomes of the animal in the peculiar kingdom of vegetation? Blessings are divided; all treasures belong not to one country alone. This luxuriant vegetation, it might be said, seems to stifle the higher life, in the animal world. Animal life is, as it were, overruled, enfeebled; it does not occupy here the first rank, which is its due; for it is dry heat, the continental element, that favors it.

From one end to the other of the animal scale, the families that seem to give to these countries their character, by the number of their species and by their relative abundance, are those connecting themselves, by their mode of life, with the aqueous, or with the vegetative element.

Hence, nothing is more splendid, more sparkling, than the insect world in South America. The inexhaustible variety of their species, the brilliancy of their colors, the size of their bodies, make them one of the most beautiful ornaments of these regions. Here live the Hercules beetle, the largest of the Coleoptera; and those brilliant broad-winged butterflies, the Menelaus, the Adonis, the Achilles, whose varying colors glitter in the sun like diamonds. But why be astonished? The existence of this little animal world is almost altogether dependent on the vegetation; the wealth of the one must create the wealth of the other.

Among the vertebrates, no family is so largely represented as that of the reptiles, for moisture is their element, and the rivers and temporary lagoons of the rainy season are peopled with Caimans, the crocodiles of the New World; the Iguanas, the most gigantic of the lizards; the Basilisks and other species, which multiply in the warm and sluggish waters. The forests harbor in great numbers those serpents of every form and figure, even to those monstrous boas, that are the terror of the natives themselves. They seem to be at home in this country.

But among the superior animals, development seems to be arrested; it is incomplete. The prevailing types are at the same time the inferior types. Among the birds, the stilt-plovers, inhabitants of the marshes and the shores, of which the number of species far surpasses, in America, that of any other continent. In the mammifera, the order of Edentata, the Armadillos, the Pangolins, the Ant-eaters, the Sloths, which characterize, more than any other family, the fauna of South America, not only in the present epoch, but also in the geological ages. And if we seek representatives of the higher orders, we find them less numerous in species smaller in size; in a word, far inferior to the corresponding animals of the Old World; in the order of the Pachyderms, instead of the elephant, the rhinoceros, the hippopotamus, those giants of the Old World, the feeble and harmless tapir and the pecari; in the Ruminants, instead of the camel and the dromedary, the lama of the Andes, which reach only half their size; instead of the lordly lion of Africa, and the ferocious tiger of the islands of the Ganges, the ounce and the jaguars of the forests of Brazil, which are scarcely more than large cats. In the monkeys, finally, those with a prehensile tail, peculiar to America, are reckoned among the least perfect, the lowest of their order in the scale of organization.

Not only are the superior animals ill represented on this continent, but they have not the strength, nor the indomitable courage, not the ferocity, nor the intelligence of the similar creatures of the Old World. In all tropical America especially, as we see, the whole animal kingdom remains in an inferior condition. It is subjected to the watery element, and to the vegetable world; for in those regions where vegetable life is the superior, animal life stands but in the second degree.

North America, however, in consequence of her more continental character, possesses some superior types, which recall to mind, and perhaps equal, those of the Old World. The majestic bison, the deer, the elk, and the bear, give evidence of that same vigorous northern nature which predominates in the temperate continents, and of which, as we shall see, North America possesses her share.

Man himself, the indigenous man I mean, bears in his whole character the ineffaceable stamp of this peculiarly vegetative nature. Living continually in the shadow of those virgin forests which overspread the land he inhabits, his whole nature has been modified thereby. The very copper hue of his complexion indicates that he lives not, like the negro, beneath the scorching sunbeams. His lymphatic temperament betrays the preponderance in his nature of the vegetative element. The Indian is of a melancholy, cold, and insensible race. "Foreign to our hopes, our joys, our griefs," says a traveller, "it is rarely that a tear moistens his eyes, or that a smile lights up his features." The most barbarous tortures cannot extort from him a single complaint, and his stoical indifference is disturbed only by vengeance or jealousy. If he sometimes exhibits a display of prodigious muscular force, he is yet without endurance. Who knows not that when the first invaders of the New World endeavored to compel the inoffensive Indians, who had received them as gods, to the rude labors of the mines and the cultivation of the soil, these men of the woods, incapable of enduring fatigue, perished in agony by thousands? And it was thereupon that the Europeans substituted for the Indian the robust and vigorous native of the Old World, the negro, who still, to this day, used as the instruments of the white man's labor, endures, I had almost said, gayly, a degree of toil equal to that which destroyed the native of the country.

The social condition of the Indian tribes is tinctured, in an equal degree, by the powerful influence of his vegetative character. The Indian has continued the man of the forest. He has seldom elevated himself above the condition of the hunter, the lowest grade on the scale of civilization. The exuberance of the soil has never been of value to him; for he asks not of the earth his nourishment. He has never even ascended to the rank of the pastoral man. With him no domestic animals are maintained to feed him with their milk, or clothe him with their fleeces, as they are by the nomadic races of the Old World. From one to the other extremity of America, we find the same lamentable spectacle; the people of the elevated table lands of Mexico and Peru are the only exception to this picture, and this exception goes far to establish the influence of the vegetative and humid nature of the lower plains of America. For if these nations do not exhibit the same character of inferiority; if they have raised themselves a little higher in the sphere of humanity, by the aid, perchance, of elements foreign to their own continent, it can be for no other cause than that, living in those heights, those aërial islands, above the influence of the hot and humid atmosphere, they have been removed from the potency of its action.

Such, gentlemen, is the order, the admirable connection of the phenomena of nature with each other. The conformation and position of the New World give to it a hot and watery climate; this impresses its own character on all the organized creation. Man himself, the one being preëminently free, is liable to its influence, in proportion as he neglects the exertion of those superior faculties wherewith he is endowed for the conquest and subjugation of that nature which was intended, not to govern, but to serve him.

We may rest, then, in this conclusion, that, as compared with the Old World, the New World is the humid side of our planet, the *oceanic*, *vegetative* world, the passive element awaiting the excitement of a livelier impulse from without. Such is the America of Nature, such was it before the arrival of the man of the Old World. We know already, and we shall see better yet hereafter, all that his superior intelligence has been enabled to effect in the way of improving upon nature.

LECTURE IX.

Geographical characteristics of the Old World — The continent of Asia-Europe — Comparison of its structure with that of America — The continental climate prevailing in the Old World — Consequences — Vegetation less abundant — Preponderance of the animal world — The Old World the country of the higher and historical races — Reciprocal action of the two worlds by means of man — Establishment of the man of the Old World in the New — Historical America compared with Europe — Alliance of the two worlds; solution of the contrast.

LADIES AND GENTLEMEN: —

The comparison we have made between the Old World and the New, and the detailed study of the first, have enabled us, I think, to determine its true character, the character assigned to it by its physical nature. The character it owes to its more oceanic position, to the abundance of the waters, to a more tropical situation, to a more fertile soil, is the marked preponderance of vegetable life over animal life. A vigorous vegetation, abundant rather than delicate, immense forests, a soil everywhere irrigated, everywhere productive — these are the wealth of America. Nature has given her all the raw materials with liberality; has lavished upon her all useful gifts.

But our globe would be incomplete, gentlemen, if this element were alone represented, if this were the only world that existed. The comparative study we have commenced has shown us already that this is not the case; the group of continents combined in the eastern hemisphere has already appeared as possessing an assemblage of characters, securing to it an entirely different nature. One of the two worlds is by no means a repetition of the other; for the Author of all things is too rich in his conceptions ever to repeat himself in his works.

We know already a good number of the physical characteristics of the Old World, an unknown world to us no more. Nevertheless, it is well to recall them here, in order to group them in a single picture, and to deduce from them the essential and characteristic feature which distinguishes it from America.

The number of the continents, double that of the New World, their grouping in a more compact and solid mass, make it already and preëminently the continental world. It is a mighty oak, with stout and sturdy trunk, while America is the slender and flexible palm tree, so dear to this continent. The Old World—if it is allowable to employ here comparisons of this nature—calls to mind the square and solid figure of man; America, the lithe shape and delicate form of woman.

The direction of the principal reliefs, the prolongation of the Old World from east to west, and its more northern position, cause it to belong rather to the temperate zones than to the zone of the tropics, and give, throughout its whole length, a more similar climate.

If America is distinguished by the simplicity of its interior structure, and by the consequent unity of character, the Old World, on the contrary, presents the variety of structure carried to its utmost limits. While America as we have seen, is constructed upon one and the same plan in the two continents, the Old World has at least three, as many as its separate masses; one for Asia and Europe, one for Africa, a third for Australia; for, in spite of their resemblance in certain general features, common to them, as the law of the reliefs has taught us, each of these three continents has none the less its special structure, which is not the same in Australia as in Africa, nor in Africa as in Asia-Europe.

The great mass of Asia-Europe, which may be well called a single continent, of a triangular form, whose western point is Europe—Asia-Europe, by itself, forms already the pendant of the two Americas. Like the New World, it is divided into two parts by a long ridge of heights, of mountain chains, and of table lands, forming a line of the highest elevations, and the axis of this continent; the Himalaya, the Hindo-Khu, the Caucasus, the Alps, the Pyrenees, are analogous to the long American Cordilleras.

This ridge also divides the Old World into two unequal parts, but is not placed on one of the edges of the continents, as in America. It is only a little out of the centre, so that it divides the whole surface into two opposite slopes, unequal certainly, but the narrower is nevertheless considerable. The northern slope is more vast: it contains all the great plains of the North, but it is less favored by the climate, and by the forms of the soil. The southern slope is less extended, but it enjoys a more beautiful climate; nature is richer there; it is more indented, more variously moulded; it possesses all those fine peninsulas, the two Indies, Arabia, Asia Minor,

Greece, Italy, Spain, which form the wealth of Asia and Europe. Figure to yourselves the coasts of the Pacific, furnished with a series of peninsulas of this description, and you will have an idea of the augmentation of wealth that would result to America from such an addition.

We will point out still another difference.

While in America the plains are always situated on the same side of the chain of the Andes, in Asia-Europe the table lands and the plains are situated on the two sides of the continental axis alternately. Thus, in Eastern Asia, the great plateau of Tubet and Mongolia is on the north, and the plains of the Ganges are on the south. In Western Asia, on the contrary, the plateaus of Afghanistan and Persia are on the south, the plains of Tartary on the north. In Europe, a different situation still; on the south of the Alps and the Pyrenees are the peninsulas and their gulfs, the mountain chains and their plateaus; in general, countries more elevated, but broken and dispersed; on the north, are chains more varied, lower; countries more continuous, less cut up, and the great plains of the North of Germany. All the combinations seem, if I may say so, to have been exhausted.

This is not all yet. The axis of Asia-Europe, instead of forming a continuous wall, without gap and without breaks, like the Andes, is composed of several isolated systems, independent of each other, often leaving wide openings between them. Sometimes it is a sea that separates them; sometimes vast plains serve as high roads to the invading nations, who pass from one side to the other of this great barrier, from the northern to the southern world. Thus, the lofty chains of the Himalaya, and of Hindo-Khu, sink towards the west and disappear, and no important ridge any longer separates the inhabitants of the steppes of Lake Aral from the dweller on the table land of Iran. The Caucasus rises abruptly from the level of the Caspian, and terminates in the same way at the margin of the Black Sea; and it is only upon the high shore of this sea that the mountains of Transylvania and the Balkan again begin to separate the northern from the southern world. This break opens to the polar winds and the northern tribes the broad gate which has cost the south such fierce assaults. The Alps, finally, do not touch the Pyrenees, and the Languedoc canal, uniting the Mediterranean with the ocean, proves the importance of this communication between the two basins.

Let us add, finally, that a large number of chains, parallel to this great line, and of others cutting it transversely, like the Bolor, the Ghauts of the Deccan, and the numerous chains of Indo-China, the Lebanus, the Oural, the Scandinavian Alps, to mention only the principal, cut the soil in all directions, divide it into a multitude of different basins, of natural regions, having their several limits, their climate, their special character, and we shall be convinced of the truth of the assertion, that in the Old World, variety of structure is carried to the extreme.

Thus, while America is distinguished by simplicity of forms and unity of plan, the Old World has in turn a diversity of reliefs, of combinations of mountain chains, of plateaus and of plains, multiplying to infinity those differences, those interior contrasts, wherewith America is less richly provided, and which, as we know, are one of the most powerful means of development.

The character of the climate of the Old World, taken as a whole, is a result of all the general features of configuration we have thus far ascertained. The vast extent of this group of continents, the height and number of its table lands, the greater elevation of its eastern regions, close it against the influences of the ocean, scarcely holding its empire over the shores. If the climate of the New World bears, in general, the oceanic character, that of the Old World, on the other hand, is dry, comparatively hot, extreme; in a word, *continental.* If the western hemisphere is the humid side of our globe, the eastern hemisphere is the arid side.

The character of the climate of the Old World is reflected in the organic beings, as we might expect, and it stamps, on the vegetation and the animal world, a special impress, important to be noticed.

In the Old World the vegetation is less universal, less plentiful, than in the two Americas. Nowhere on its surface do we encounter virgin forests, whose extent can be compared with the boundless selvas of the Amazon; they are found, doubtless, in the tropical regions of Africa and the Indies; but they are rather local phenomena, and do not give their character to vast countries. On the other hand, the Old World is the world of steppes and deserts. Nowhere else are those dry and barren plains so numerous, so extensive, so unbroken. It is enough to mention the boundless steppes of Russia and Caucasus, of Siberia and the Altai, of Tartary and of

Turkestan; to recall to mind the great zone of deserts obliquely traversing the Old World in its greatest length, from the shores of the Atlantic, through Sahara, Arabia, Eastern Persia, and Mongolia, to the Pacific Ocean, and occupying all the central part of the lands of the three continents united, to be convinced that the distinctive character of the climate of the Old World is dryness.

The general forms and aspect of the plants themselves, at once declare the parsimony wherewith nature has provided for them the moisture so essential to their full development. Instead of expanding their surface for evaporation and absorption, their leaves seem to fold upon themselves, to concentrate themselves into a smaller volume; they have a tendency to approach the linear shape, the pointed form we notice in the pines; they often become membranous, leathery; or the plant is covered with a soft down, with a nap, or even with prickles, which are only leaves or branchlets, transformed and hardened under the influence of a dry air. Or, still further, they take those plump, fleshy, cylindrical forms, which seem struggling to contain the greatest quantity of vegetable matter in the smallest possible volume. Such is the flora of Southern Africa, with its Stapelias, its juicy Mesembryanthems, its brilliant Aloes, its delicate Mimosas, its Metrosideri, its heaths without number. Such is that of Australia with its forests of Eucalypti, its Banksias, and its Casuarinas, with their long and naked, pendant, thread-like branches. Such, moreover, is the flora of the steppes and the deserts of Arabia and Gobi, (for there, also, are still some scanty representatives of the vegetable world,) consisting entirely of plants of a dry and ligneous nature, often clothed with white down, or of gray hues, imitating the color of the dust of the desert. In all these countries the forests are rare, of small extent, of little density; the scattered trees are not invaded by those woody climbers which elsewhere entangle and interweave them, and form those impenetrable masses of verdure which characterize the tropical wilds of the New World. Thus, in the vegetable kingdom of the Old World, there is oftener a scarcity, oftener a sufficiency, but rarely an abundance.

Nevertheless, because the vegetable does not reign there by its mass, is this saying that it reaches a less perfect organization? No, gentlemen, no such thing. It is this dry and warm climate that produces the delicate fruits of Persia and of Asia Minor, elaborates those refined juices, those perfumes, those fine aromas of the East, the fame of which was already established in the remotest antiquity. These same regions of the Old World have given us coffee from Arabia, and tea from the uttermost Orient, so precious to all the civilized nations. The East Indies and their archipelago, as we have said, under the influence of the vigor of the continent and the moisture of the ocean, yield those concentrated products, those strong spices, the nutmeg, the clove, the ginger, there in its native country; the pepper, the cinnamon, of which the whole world makes use. It is these same countries that present us the largest leaves and flowers known; there, also, grows the Banyan tree, the symbol of vegetable strength; it is in Africa that the huge Baobab unfolds itself — the Adansonia, whose trunk sometimes measures twenty-five feet in diameter. But, gentlemen, let us acknowledge it, these are the products of favored spots; the common rule, — I do not speak of quality, but of quantity, — the general rule of the Old World is economy, and not superfluity.

But if vegetation in the Old World seems reduced to a subordinate place, it is not so with animal life; this we find here in its fulness, and varied to a degree the New World knows nothing of. While it seems to be impoverished and subdued in the dank atmosphere of tropical America, it flourishes most of all in the dry, hot, exciting climate of Africa and Asia.

America, as we have seen, is the domain of insects and reptiles, which prosper in the humid and vegetative element.

The Old World is the domain of the higher animals, the Mammalia, the number of which, their variety, their strength, and their size, give character to the animal kingdom of these countries. As we have said, when comparing the animal kingdoms of the two worlds, not only the representatives of the corresponding families are larger and stronger in the Old World, but they appear in more numerous genera and more varied species, and even exhibit types entirely foreign to America, as the giraffe, the giant of the Ruminants.

The servants and companions of man, the horse, that noble animal which he can neither forego for his wants nor for his pleasures; the ox, more useful still; the dog, his faithful friend, are gifts the Old World has bestowed upon the New. Finally, the presence of the Chimpanzi of Africa, and of the Orang-outang of the Indies, whose figure comes the nearest to man's, proves that the development of animal life reaches here the utmost limits it can attain, apart from man himself.

The animal kingdom, as we see, has, in the Old World, the preponderance over the vegetable, as in the New World the vegetable had over the animal. The kind of enemies man encounters in the one and the other world, when he struggles with nature, already tell us what is their character. In America, the overflowing rivers and their lowlands; the virgin forest, the climbing plants and their impenetrable thickets; the poisonous reptiles, and the devouring insects multiplying there, are his most formidable enemies. In Africa or in Asia, it is thirst, the moving sands, and the burning heat of the sun; it is the lion, the tiger, the hyena, and all the ferocious inhabitants of the desert, that menace his life and encompass him with ever-recurring dangers.

Let us raise ourselves higher still, and pass into the province of man himself. We find here the contrast between the two worlds still more marked. Instead of a single race, the copper-colored, dwelling in both Americas, from Labrador to Cape Horn, four different races, if not five, belong to the Old World, and testify to the variety of its plastic forms, and to their powerful influence upon the organization of man.

The white race is distinguished above them all: the most perfect type of humanity; the race best endowed with the gifts of intelligence, and with the profound moral and religious sentiment that brings man near to Him of whom he is the earthly image. To this race belong, without exception, all the nations of high civilization, the truly historical nations; this still represents the highest degree of progress attained by mankind. After it, the Mongolian and Malayan races, which might be called the semi-historical nations, are still superior in civilization to the copper-colored. If we take even those races of the temperate regions of the Old World, at the lowest degree of the social scale, the nomadic tribes of the plateaus of Eastern Asia and of the western steppes, they are still far superior to the hunting tribes of the two Americas. There is even in the tropical man of the Old World, in Africa at least, a somewhat of native vigor of vital energy, manifested by his sanguine temperament, by his gayety, by his lively affections, and by his muscular strength, placing him higher than the Indian of tropical America. His social state, even, has made a step in advance. The negro tribes of Congo and Soudan form real commonwealths; they are acquainted with agriculture.

The density of the human population in the two worlds speaks with still greater emphasis. Taking the least uncertain numbers, we find that, while Europe counts 89 inhabitants to the square mile, Asia 32, and Africa 14, America has only 4. If we compare, then, either the races of the two worlds, or the civilized nations of Europe and of Asia, with those of the ancient inhabitants of Mexico and Peru; or, finally, the least cultivated tribes, the nomads and the negroes, with the hunting Indians of North and South America, the advantage, in various respects, will remain with the Old World.

Thus the Old and the New World are distinguished by an assemblage of different and opposite characterstics, making a separate type of each, and giving to each a peculiar physical aspect. In one, the simplicity of forms, the moist climate, the dominance of vegetation, declare the passive element; in the other, the variety of forms, the dry and extreme climate, the animal world, all proclaim the active element. They are opposed to each other as the vegetable and the animal.

Furthermore, in this great contrast, as in that of land and water, we find an inequality between the two factors, considering them as we are now doing, with respect to their physical nature; one of the two worlds appears to us as the superior, the other as the inferior. The Old World holds the first rank by its mass, by the number of its continents, by the variety and richness of its structure, by its continental climate opposed to the oceanic; by the preponderance of animal over vegetable life, by the number and superiority of its races of men; finally, it is the primitive seat of high civilization.

But these two parts of our planet are only the parts of one organic whole. We see in advance—and the law we have so often traced out in the course of our studies confirms it—we see in advance, that two individuals so different from each other cannot be confronted without entering into relations, without commencing a life of interchanges, that will enrich them both. Thence, by this mutual action of the two worlds, will be unfolded all of the wealth of life, the germs whereof are deposited in each; a grand unity will be constituted, which, embracing both, will raise them to the highest degree of improvement Providence seems to have decreed for them by nature itself, but which they cannot attain without each other's aid.

The inequality we have just established, is, then, an additional source of wealth, for it summons forth the mutual actions, and hastens the solution of this great contrast.

But how will this mutual action take place? Physical nature has exhausted its means of action for producing, by the intervention of the atmosphere and the winds, the mixture of land and sea, the diversity of climates and of organized beings, making the two worlds two distinct individuals; it cannot go further; it belongs now to man, for whom they were made, to human societies, to continue this work, to blend their two natures, and to establish between the two worlds the permanent bond that is to unite them. In their action upon the peoples and nations of the globe, all their physical differences must be reflected, and may be expected to display their true importance.

America lies glutted with its vegetable wealth, unworked, solitary. Its immense forests, its savannas, every year cover its soil with their remains, which, accumulated during the long ages of the world, form that deep bed of vegetable mould, that precious soil, awaiting only the hand of man to work out all the wealth of its inexhaustible fertility. Meantime, the human race of the New World, the Indian, the primitive owner of these vast territories, shows himself incapable or careless of the work; never has he opened the soil with his ploughshare, to demand the treasures it encloses. Hunting is his livelihood; war his holiday. Upon a soil able to support millions of men in plenty, a few scattered inhabitants lead a wretched existence in the bosom of the wilderness.

Side by side with so much unused wealth, see the Old World, exhausted by long cultivation, overloaded with an exuberant population, full of spirit and life, but to whom severe labor hardly gives subsistence for the day; devoured by activity, but wanting resources and space to expand itself; and you will perceive that this state of things, that a disproportion so startling, cannot long exist. The gifts God bestows on man He requires should be employed, and He takes from him who does not put it to use, the talent which has been entrusted to him.

As the plant is made for the animal, as the vegetable world is made for the animal world, America is made for the man of the Old World. It is to the latter, as the active principle, that the first onset belongs.

Everything in nature is admirably prepared for this great work. The two worlds are looking face to face, and are, as it were, inclining towards each other. The Old World bends towards the New, and is ready to pour out its tribes, whom a resistless descent of the reliefs seems to sweep towards the Atlantic. America looks towards the Old World; all its slopes and its long plains sweep down to the Atlantic, towards Europe. It seems to wait with open and eager arms the beneficent influence of the man of the Old World. No barrier opposes their progress; the Andes and the Rocky

Mountains, banished to the other shore of the continent, will place no obstacle in their path. Soon the moment will come.

The man of the Old World sets out upon his way. Leaving the highlands of Asia, he descends from station to station towards Europe. Each of his steps is marked by a new civilization superior to the preceding, by a greater power of development. Arrived at the Atlantic, he pauses on the shore of this unknown ocean, the bounds of which he knows not, and turns upon his footprints for an instant. Under the influence of the soil of Europe, so richly organized, he works out slowly the numerous germs wherewith he is endowed. After this long and teeming repose, his faculties are reäwakened, he is reänimated. At the close of the fifteenth century, an unaccustomed movement agitates and vexes him from one end of the continent to the other. He has tilled the impoverished soil, and yet the number of his offspring increases. He turns his looks at once towards the east and the west, and sets out in search of new countries. His horizon enlarges; his activity preys upon him; he breaks his bounds.

Then recommences his adventurous career westward, as in the earliest ages. His intelligence has grown, and with it his power and hardihood. Under the guidance of the genius of the age, he attacks this dreaded ocean, of which, to this time, he knows only the margin. He abandons himself to the winds and the currents, which bear him gently towards the coasts of America. He is enraptured as he treads the shore of this land of wonders, still more adorned in his eyes by all the fascinations his ardent imagination lends it.

The European establishes himself little by little upon this new land; he gets a foot-hold but slowly; for, to his shame be it said, the thirst for gold seems the only motive urging him thither; for gold, that factitious, cheating, transitory wealth, which in the long run impoverishes him who has it, because it puts his faculties asleep; that gold, fatal to Spain, the abundant possession of which was the signal of her decline. To make a fortune rapidly, by all possible means, and to return to Europe to enjoy it, this was the aim of the earliest colonists. These are not the true laborers in the great work that is beginning; these are only the trappers; these are not the civilizers of the New World; not to them shall it be given to be its true possessors.

Meantime new bands from beyond the seas soon discover that the real wealth of America lies in the fertility of the soil. Then begin the interchanges. The European plants, in this still virgin land, the useful vegetables he brings from the Old World, the sugar-cane, the coffee, the cotton, the spices, the cereal grains, more precious still, and draws therefrom abundant harvests. The New World gives to Europe, in exchange, the cocoa, the vanilla, the quinquina, the potato, above all, alone worth all the rest. The domestic animals, which are wanting in America, follow the footsteps of the colonists thither; the horse, the ass, the ox, the swine, all these useful companions of man, that act so important a part in the domestic economy of civilized nations, henceforth enrich this second half of the earth.

For a long time America is a daughter of the Old World, in her minority; and nevertheless, the colonial system already reacts profoundly upon the development of the European nations. During the three centuries following the discovery, the questions connected with the commerce of the world and the possession of the colonies grow every moment in importance. Every day brings with it the establishment of new colonies, and augments and reinforces those already existing. A local life makes no delay in displaying itself on this fresh soil. Whole peoples take root and increase with rapidity in the midst of that nature which yields them everything in abundance. They ask no more help from the mother country; they are in a condition to furnish it to *her;* the consciousness of strength grows with their prosperity.

But the hour of independence has struck; the fruit is ripe; it drops from the tree. The sons of the Old World have adopted America for their country; she has become their beloved mother. America takes her position face to face with Europe, not as a minor, but as a full-aged daughter—free, for it is her right. She throws herself alone, and on her own account, released from guardianship, with demeanor more open, more frank, more rapid, into the career of civilization. Now commences a new antagonism, more serious, more vast in its proportions. The two worlds treat as power with power for two free and independent beings look upon each other. But, gentlemen, they are not enemies;

they are too well adapted, too truly made for each other; they have too much need of each other; they are too much the complement of each other, not to unite for their common interest. Their differences will only serve to excite a more active life, a more extensive and lasting interchange of all that each can give in abundance to its rival.

Here, in fact, we find all the elements, all the conditions, of a well-assorted union, a true marriage. Is there not between the peoples of the two worlds a common basis, an essential, indissoluble tie, which they are not at liberty to break? Are they not all the children of a common mother-race? the offspring of the same civilization, the worshippers of the same one God and Saviour? And yet there is an individual difference of character between them, arising essentially from the special work to which each seems to have been called as to an appointed task. For the American, this task is to work the virgin soil, and the wealth of the land Providence has granted to him, for his own benefit and that of the whole world. For this is the first work to be done, that whereon the futurity of America depends. He is accomplishing it—who does not know?—with a fiery activity. He has not too much of all the resources of industry the Old World and his own experience place at his disposal, to subdue and fashion at his pleasure this still somewhat savage nature.

Agriculture here already assumes proportions unknown everywhere else. Commerce, whose business is helping the world to share the wealth of its soil, is carried on upon the most magnificent scale, and cannot but become still more extended. From the very centre of the oceans where she reposes, America sends her ships and her merchandise to the ends of the earth. Steam will soon join the shores of the Atlantic and the Pacific, and place the United States on the great highway from Europe to China. Thus the American displays in every way a spirit of enterprise that goes even to the length of audacity. Nothing daunts him, nothing seems impossible to his activity. Every brain is teeming with the most gigantic projects, which find always an echo and support. There is certainly something of grandeur in the spectacle of the youthful vigor the inhabitant of the New World displays, of the intelligent energy with which he does his work. Whatever be its object — were it even not the most exalted — still such energy is worthy of admiration.

The work of Europe, her special task, at the present moment, is not the same; for her position is altogether different. Without doubt, industry, commerce, agriculture, employ a large part of her activity; but the exercise of these arts is not the predominant and characteristic feature of that ancient society. Other cares occupy her. The desire to know, rather than to possess; reflection, more than action; science, more than its application; movement and activity in the intellectual and moral world, rather than in the material world; — these are what distinguish the Old World and its ancient civilization.

Thus it is there that the high philosophical, moral, social questions are treated, which so profoundly task the present age; it is there that the thousand ideas, the thousand diverse systems in all the branches of human science, whose variety seems an inextricable confusion to the eyes of the mind that does not master them, bud and blossom. The European is the man of ripened age, who reflects upon men and things, analyzes the causes, and seeks to understand the lessons of the spectacle the world presents. The American is the young man, full of fire and energy, surrendering himself entirely to his activity, and drawing from life the practical shrewdness and the sound sense experience gives. While the European discusses and reasons, the American acts and executes.

But, gentlemen, who does not see what there is exclusive in these two tendencies? Who does not understand to how many mutual wants these differences must give birth? how many exchanges of every kind they must stimulate? what activity, what fulness of life and of growth for both, what perfection of both will be the result of these intimate relations? Distressed Europe seemed unable to live longer without emptying its population into the lap of America. America cannot fulfil her destinies unless wrought out and brought into use by the intelligent races of the Old World. When this work, now just commencing, shall have been finished, then only shall we know all the importance of America to the entire race of man; all the importance of the reactions it is summoned to exercise upon the Old World.

The Old World is the world of germs; the New, the fruitful bosom giving them increase. Europe thinks; America acts. All these differences, calling for and completing each other, are they not the promise that this contrast of the Old World and the New will soon also be resolved into a grand and beautiful harmony that shall embrace the whole earth?

Yes, gentlemen, the futurity and the prosperity of mankind depend on the union of the two worlds. The bridals have been solemnized. We have witnessed the first interview, the preliminaries, the betrothal, the espousal, so fortunate for both. We already see enough to authorize us to cherish the fairest hopes, and to expect with confidence their realization.

LECTURE X.

Contrast of the three continents of the North and the three continents of the South — Physical characteristics of the two groups; the former more articulated, more consolidated, more similar; the latter more entire, more isolated, more different — These differences and analogies reproduced in the vegetation and the animal world — The three continents of the North temperate; the three of the South tropical — Superiority of the tropical climate in nature — Gradual increase of life, of the variety and improvement of the types of organized beings, in proportion to the warmth, from the poles to the equatorial regions — Man alone forms an exception — Law of the distribution of the human races — Geographical centre of mankind marked by the race of the highest perfection — Gradual degeneracy of the human type towards the southern extremities of the continents — The geographical distribution of the races of man and the animals founded upon a different principle — Advantage of the temperate climate for the improvement of man.

LADIES AND GENTLEMEN: —

We have considered the whole of the terrestrial masses, as grouped in two great individuals, the Old and the New World, which have exhibited themselves as possessing each a special character, particular advantages, but completing each other, and forming, as it were, two halves of one great organization. The union of these two worlds, the resources of which have been drawn out by the most intelligent and active races, by the most advanced societies, is the condition, and must become the means of a progress of the human race, much superior to what it would have been in each of the two worlds isolated from each other. It is at least the hope given us by the law of the resolution of the contrasts into a more perfect harmonious unity, which is the natural product of all normal development. All that passes under our eyes at the present day, this life of interchanges between the two worlds, so active, so powerful, so progressive, is a proof that we are advancing irresistibly to so desirable a result.

We have now to consider our continents only under one remaining aspect, under a third point of view, which I hope will disclose to us some of the hidden influences they seem to exercise upon the life of man; or rather, which will enable us to observe one of those admirable harmonies of nature and history, arranged by the Creator himself for the improvement of this privileged being, for whom all nature seems made.

This third contrast, we have said, is that of the northern and of the southern hemisphere, or rather of the three continents of the North and the three continents of the South. We have considered the earth as divided into an Eastern World and a Western World; we shall now see it distributed into a Northern World and a Southern World.

I recall this curious disposition of the continental masses I pointed out, at our first interview, according to Steffens, which consists in this, that each northern continent has south of it a southern continent, more or less connected with it, whether materially by an isthmus or a chain of islands and an archipelago, or by the proximity of their extreme lands. The two continents, thus brought near together, make always a pair, the individuals being at once connected and opposite in nature. Such are the two Americas, — a perfect type of the kind; such, again, are Europe and Africa, Asia and Australia.

This arrangement, then, gives us three continents in the North, and three in the South. Now, combining in this point of view the prevailing features of their physical configuration, of their situation and of their climate, I would show you: —

1. That the continents composing each of the two groups have common characters, the three in the North resembling each other, and the three in the South presenting equally strong analogies.

2. That these characters are different and opposed in the two groups, and constitute a contrast.

3. That this dissimilar nature assigns to the one a very different part from the others in the progress of human society.

Let us see, first, how they are distinguished by their general forms and by their configuration.

The continents of the North are more outspread, more extended, and, taken together, much more vast. They embrace all the plains of the arctic and temperate regions, the most considerable and the most continuous on the surface of the globe, and forming a great circular zone of low lands around the Frozen Ocean. The southern continents are more contracted, more narrowed and pointed, and, in the whole, less extensive. While the three first embrace a surface of 22½ millions of square miles, the last comprise only 16⅓ millions.

The continents of the North are more indented, more

articulated; their contours are more varied. Gulfs and inland seas cut very deep into the mass of their lands, and detach from the principal trunk a multitude of peninsulas, which, like so many different organs and members, are prepared for a life, in some sort, independent. A great number of continental islands are scattered along their shores, and are a new source of wealth to them. The plastic forms of the soil are still more varied. We have already seen that, in this respect, Europe and Asia present the most complicated structure, and the relative situations of the mountain chains and of their plateaus and their plains, exhaust, so to speak, all the possible combinations.

The southern continents, on the other hand, are massive, entire, without indentations, without inland seas or deep inlets, scanty in articulations of every kind, and in islands. They are trunks without members, bodies without organs, and the simplicity of their interior structure answers to the poverty of their exterior forms.

These differences are carried to the extreme in the Old World, where the rich border of peninsulas which deck the South of Asia and of Europe, hanging like the ample folds or the fringes of a royal robe, form a striking contrast to the mean and naked lines of Africa and Australia. In the New World, where this contrast is softened, by reason of the unity of plan we have clearly made out, the northern continent meantime possesses by itself the few peninsulas which detach themselves from its coasts, Yucatan, Florida, Nova Scotia, Labrador, California and the peninsula of Aliaska.

Already, then, by the forms of their contours and of their relief, the continents of the North are more open to maritime life, to the life of commerce: they are more richly organized; they are better made to stimulate improvement.

The relative situations of the continents of the two groups are equally dissimilar.

The northern continents are brought nearer together, more consolidated. United, they form the central mass of all the lands of the globe, whence the others appear to radiate in all directions, losing themselves as they taper off in the ocean. For this reason, they have a more continental character. Owing to this greater nearness, to the facility of communication between one continent and another, to the analogy of their climate, which we shall speak of by-and-by, the three northern continents have a mutual relationship not to be mistaken. From the shores of the Pacific Ocean, along the coast of temperate Asia, even to the western extremity of Europe, the vegetation presents the same aspect, the same general physiognomy. The European traveller finds, from one end to the other of this immense space, the pine forests, the oaks, the elms, the maples his eye has been accustomed to from infancy. In the Himalaya, the Caucasus, or the Balkan, he beholds again with delight those humble but graceful forms of the flora he has become acquainted with in the Alps and the Pyrenees. If he crosses the Atlantic, what surprises him at the first glance is, not the novelty of the vegetable forms which he was perhaps expecting after a voyage of thirty-five hundred miles; it is the resemblance of physiognomy and aspect, so great that, in the bosom of the vast forests of Ohio or Canada, he might almost believe he had not quitted the soil of Europe. Nevertheless, we hasten to say, this resemblance is not identity. The eye of the botanist, even that of the simple observer, would soon perceive that, if the types remain the same, the species are different. If they are almost always analogous, they are seldom identical.

In the animal world, the same analogy still. Nothing is more alike, at the first view, for example, than those thousands of coleopterous insects, which inhabit the two worlds. The same air, the same look in the corresponding species. It is a singular fact, observed also in the vegetative kingdom, and still a mystery, that a given genus, in Europe composed of a determinate number of species, is found again in America, with an almost equal number of corresponding species, with the same particularities of forms repeated, even to the design and to the same disposition of colors. And yet, to the trained eye of the naturalist, every American species constitutes one very distinct from the analogous species of the European continent. What takes place with regard to the genera and species is further true of certain families and tribes. The examples of this are numerous, and I would cite them but for the fear of offending your ears by names which would appear barbarous. Relations of the same kind exist between the vertebrates, between the fishes, between the birds of the two worlds, and it is to these deceptive resemblances that the confusion of species and the mistakes of synonyms are owing, so numerous in American zoölogy, and so hurtful to its progress. The mammalia, finally, make no exception to this law. The reindeer is common to the polar regions of the three continents; the bison reminds one of the wild bull and the ox of Europe and of Asia; the bears, also, are but slightly different from those of the Old World; the elk and certain kinds of stags are so similar, that the zoölogist is still in doubt whether they constitute different species or not.

Thus the resemblance of the organized beings in the three continents of the North is one of their distinctive characters; and this character is due to the circumstance that, in proportion as the species change with the longitude, their place is taken generally, not by new types, but by analogous species. Doubtless the similarity of climate is one of the most active causes of this resemblance; for the variety of the genera, the differences between the species of the three continents, augment according to the elevation of temperature; but this is not enough to explain the fact entirely; we shall see that the continents of the South, under similar latitudes and in similar temperatures, offer types of animals and of vegetation very different in each of them.

The continents of the South are more remote from each other than the foregoing. Broad oceans separate them even to isolation. Scarcely any communication is possible between lands so distant; at all events, they are only indirect. Shut up in themselves, incapable of reacting upon each other and of modifying their respective natures, these continents are excluded from all community of life. What is there astonishing, then, in seeing their differences carried to an extreme, their characters exaggerated?

The organized beings of the two kingdoms in these three continents have, in reality, almost ceased to possess anything in common. Not only the species that characterize their floras and their faunas are different, but they are no longer analogous, and the prevailing forms, the grand types, are partly quite different. This is true, above all, of their southern extremities, of their points, more isolated still than the central or northern parts. In Australia, it is the gigantic Myrtaceæ, the flaring Eucalypti, so varied; the Melaleucas, the numerous species of which compose the greater part of the trees of the forests; it is the graceful Mimosas, and their Acacias with leaf-like branches; it is the meagre Casuarinas, and still other forms whose stunted foliage betrays the dryness of the soil, that give a particular physiognomy to the whole aspect of nature. Marsupials of huge size, the kangaroos, and other analogous animals, gambol in these forests and in the vast savannas; in the marshes, the Ornithorhynchus, unknown to every other continent, whose shapeless type brings to mind the earliest ages of the world, and seems not to belong to the existing epoch.

In Southern Africa, other forms are found, another nature. With the pale foliage of the Proteaceæ, are blended the Stapelias, the aloes, with their pulpy leaves and their brilliant corollas; the Irideæ, with their bold bearing and splendid colors; the geraniums; the heaths, above all, the numberless species of which astonish the eye by their variety, as much as they charm it by their modest grace. In place of the jumping kangaroos, herds of nimble gazelles and graceful antelopes wander over the vast plains of the high regions of Africa. The hyena, the panther, the lion, strangers to Australia, and witnesses to a stronger and nobler nature; the giraffe, which Africa alone possesses; all, in a word, assume another aspect and a peculiar stamp.

If we pass now to South America, the animated world changes its physiognomy still again. The preponderance and the variety of its palm trees; in the richest regions, the cactus, whose heavy form contrasts with the dazzling colors of the flowers; then the clumsy armadillos, the tapirs, the ant-eaters, the long-tailed apes, and so many other animals characteristic of this continent, which we have already named;—all this has nothing to remind us of Africa.

Thus, between the three southern continents there is no community; out of 437 genera of the Australian flora, scarcely 80 are met with in Africa; no analogous species, substitutes for each other; none of those social plants covering whole provinces in Europe, Asia, and North America, and giving them a like character. The 280 species of heath of the Cape occupy a space scarcely so extensive as is occupied in the North of Europe by the common heather (*Erica vulgaris*) alone, so extensively growing in its barren regions. Thus, in the North, we have combination, association, resemblance; in the South, separation, isolation, dissimilarity.

But if the northern continents are evidently favored by their forms and their grouping, is it the same also with their climate?

The astronomical situation of these two groups is, in reality, quite different. In consequence of the general arrangement of the lands, crowding them in a mass towards the North, the three continents of the North are situated almost entirely in the temperate zone, in the middle latitudes. North America and Europe are entirely in the temperate and frozen zones; Asia is so with respect to its principal mass, and touches the tropical regions only by its southern extremity. Thus it is seven parts temperate and cold for one tropical.

The southern continents, on the contrary, expose their principal and most important mass to the rays of the equatorial sun. Africa has four parts out of five in the tropical zone, and the fifth is situated in the warm temperate, and moreover is divided into two narrow belts, separated on the north and the south of the body of the continent. South America has five parts out of six in the tropics, and the sixth part, temperate, is composed only of the southern point, the poorest in all respects, which cannot claim to stamp its character. Australia, finally, belongs three fifths only to the torrid zone; nevertheless, it should be said that the other two fifths, situated in the warm temperate zone, give it its distinctive physiognomy, so that we have called it the subtropical continent.

Thus, taken as a whole, and in their prevailing character, the three northern continents are temperate, the three southern continents are tropical.

Which are the most favored? which are those we consider superior to the others?

The answer would be easy, if the existence of the

continents had no other definite end than the exhibition of the whole physical life of nature. But let us not forget that they are to serve a much higher end still; they are to serve the development of man, and of human societies. It is in this two-fold point of view that we ought to consider them.

In the order of nature, and at the first approach, we cannot deny to the tropical continents a marked superiority. The most powerful spring of physical life, the most active source, surpassing all the others, is the heat of that life-giving orb the ancient poets sang, and the nations of the world, forgetting the only true Creator of all things, adored as the parent of Nature. But by reason of the spherical form of the earth, each district of the surface receives an unequal portion. Slanting, scattered, and feeble in the regions neighboring the poles, the beams of the sun assume more strength, and fall thicker in the middle regions; in those of the equator only, they gain all their intensity, all their splendor. Now, in this same proportion, we see the development of life increase in energy and variety, from the poles to the equator.

What do we, in reality, see in the polar and frozen countries of the North of our continent? During the greatest part of the year, life seems almost extinguished by the rigorous cold of a perpetual winter. A colorless and stunted vegetation, a few creeping shrubs, none of those stately forests, which everywhere make the ornament of the landscape; endless plains covered with mosses and lichens, composed of only a few species, notwithstanding the immense number of their individuals. This is the flora of the cold regions. The preponderance of the cryptogamous plants, that is, of the inferior forms of vegetation, the small number of the genera and the species, the absence or scarcity of arborescent vegetation, give it that character of poverty and uniformity which strikes us in these desolated lands. The animal kingdom, thanks to greater freedom of locomotion, is better represented; but the small number of types and the preponderance of marine animals, still keep up a character of inferiority not to be misunderstood.

In the temperate zone, the number of genera and species is more than doubled; the superior types acquire a fuller development and more importance. In the vegetation, the preponderance of the phanerogamous plants, the beauty of the forests, the appearance of evergreen trees, are the signs of an immense progress. Meantime, the soft tints, the modest forms, the winter sleep, still interrupting the life of vegetation during long months, tell us that the triumph of life is not yet complete. The same progress goes on in animal life; the land animals prevail; the animal species become more numerous and more diverse.

But it is in the hot region of the tropics that the life of nature displays its fullest energy, its greatest diversity, its most dazzling splendors. We have already seen what it can produce in those favored countries of India and the Indian archipelago, where all the conditions seem brought together to secure to physical life its richest development. The cryptogamous plants attain, in the arborescent ferns, the proportions of our forest trees.

The grasses, which we only know in our climates under the humble forms they put on in our fields and pastures, rise into the elegant and majestic bamboo, to the height of 60 to 70 feet, and become real trees, whose hard and hollow trunks serve for the construction of public edifices, as well as for that of private houses. There, the entire forests seem double in height, and of a density unknown in our climates. A single tree is a garden, wherein a hundred different plants intertwine their branches, and display their brilliant flowers on a ground of verdure, where the varied hues and forms of their leaves are blended together. The number of the species, the beauty of the types, are not less astonishing. While, in America, the temperate zones of the two hemispheres furnish scarcely more than four thousand species of plants, the tropical region of this same continent has already made known more than thirteen thousand; so that probably the comparatively narrow zone of the tropics contains much more than half of the vegetable species living on the surface of our continents.

The animal kingdom is no less developed, as we already know, in this privileged zone. The boundless variety of species, the vivacity of the colors, the diversity of the shades, strike us in the insects and the birds. We admire the lofty stature and the strength of those great pachyderms that people its forests and its rivers; the force and vigor of the ferocious inhabitants of the deserts of Africa and the Ganges. It is true, gentlemen; here Nature triumphs; here she displays herself in all her brilliancy.

Such is the law in the physical world. Nature goes on adding perfection to perfection, from the polar regions to the temperate zones, from the temperate zones to the region of the greatest heat. Animal life grows in strength and development; the types are improved; intelligence increases; the forms approach the human figure; the orang-outang already stands erect upon his feet; trained up by man, he has been seen to sit at his table and to eat with him; the negro of the woods, deceived by these appearances, regards him as a degenerated brother, who holds his tongue only from a desire to get rid of work. Evidently the development of the animal here touches upon its highest expression.

This ascending series will then rise to its termination in man, who, in his figure, is the crowning excellence of the whole animal world, and the realization of its very idea; and the tropical man also will be the highest, the purest type of humanity, and, physically speaking, the most beautiful of his species. All zoölogy, all nature, gentlemen, authorize us to draw this conclusion, and, for my part, I have no hesitation in believing that it would be so if man had no other rank upon this earth and no other functions than those assigned him by his physical nature.

No, indeed, gentlemen, it is no such thing. Who does not know that man makes here a wonderful exception? Far from exhibiting that harmonious outline, those noble and elevated forms, all those perfections the chisel of a Phidias or a Praxiteles has combined upon a single head, the tropical man displays only those unfortunate figures which seem to approach ever nearer and nearer the animal, and which betray the instincts of the brute those figures which we always behold with, I know not what of secret uneasiness, that would threaten to grow into disgust, were not the feeling lost in pity still more profound, and in the charity of a Christian heart. Even in that India, where physical life attains the utmost limits known to our earth, the indigenous man is a black; the white race—history compels us to believe it—has descended thither from the temperate regions of Western Asia.

If the distribution of the human races on the surface of the globe does not follow the law of the rest of nature, what, then, is the law that regulates it? Or, indeed, is there some great fact which may prove to be a rule in this seeming confusion?

Much has been said, gentlemen, much has been written, on this important question of the human races—one of the most difficult and the most delicate the science of nature and history can propose to itself. I am not going to discuss it here; but what I desire is, to establish, in this province also, a great general fact, which, as it seems to me, has not been sufficiently insisted on, and to which has not been attributed the importance it deserves. This fact is the following:

While all the types of animals and of plants go on decreasing in perfection, from the equatorial to the polar regions, in proportion to the temperatures, man presents to our view his purest, his most perfect type, at the very centre of the temperate continents, at the centre of Asia-Europe, in the regions of Iran, of Armenia, and of the Caucasus; and, departing from this geographical centre in the three grand directions of the lands, the types gradually lose the beauty of their forms, in proportion to their distance, even to the extreme points of the southern continents, where we find the most deformed and degenerate races, and the lowest in the scale of humanity.

Please to cast a glance upon this series of drawings, portraits taken from nature, from individuals who seemed to have the most characteristic features of their respective races, and you will be convinced that the fact exists. (See plates v. and vi.)

Let us take for a type of the central region of Western Asia, this head of a Caucasian. What strikes us immediately is the regularity of the features, the grace of the lines, the perfect harmony of all the figure. The head is oval; no part is too prominent beyond the others; nothing salient nor angular disturbs the softness of the lines that round it. The face is divided into three equal parts by the line of the eyes and that of the mouth. The eyes are large, well cut, not too near the nose nor too far from it; their axis is placed on a single straight line, at right angles with the line of the nose. The facial angle is 90 degrees. The stature is tall, lithe, well proportioned; the shoulders neither too broad nor too narrow. The length of the extended arms is equal to the whole height of the body; in one word, all the proportions reveal the perfect harmony which is the essence of beauty. Such is the type of the white race — the Caucasian, as it has been agreed to call it — the most pure, the most perfect type of humanity.

In proportion as we depart from the geographical centre of the races of man, the regularity diminishes, the harmony of the proportions disappears. Let us follow them first in the direction of Europe and of Africa.

Although the European may be considered as making a part of this central race, his features have less of regularity, of symmetry; but more animation, more mobility, more life, more expression. In him, beauty is less physical and more moral.

If we pass into Africa, we meet the Arab, who, whether in his own country or in Algeria, shows already a forehead slightly retreating, a head lengthened out of proportion. The Galla of Abyssinia is almost black, his long hair begins to crisp, his lips are often thick. The Caffre has the woolly hair and thick lips of the negro. The Hottentot, lastly, so struck the first colonists of the Cape by his ugliness, that he served for a long time as a symbol to express the most degraded state of humanity. On the other coast of Africa, more remote from Asia, the degeneracy of form is still more rapid. The Berbers of the Atlas still evidently belong to the Caucasian race; but their prolonged head, a tendency in the mouth to pouting, the spare and meagre forms, a deeper color, already herald a marked degeneration. The Fellatahs of Soudan, and still more the inhabitants of Senegal, bring us to the pure type of the Congo negro. In the latter, the retreating forehead, the prominent mouth, the thick lips, the flat nose, the woolly head, the strongly developed hind-head, announce the preponderance of the sensual and physical appetites over the nobler faculties of the intellect. At the extremity of Africa, the miserable Bushmen are still lower than the Hottentots; and, placed by the side of the

Plate V Portrait Types of the different Races of Men (See Plate VI.)

Plate VI.

Caucasian, make us see how immense the distance which separates them.

If, turning towards Eastern Asia, we direct our looks as far as the extremity of Australia, the decreasing beauty of the form is not less perceptible, not less gradual. The Mongolian, with his prominent cheek-bones, his eyes compressed, wide apart, and elevated at their outer corners, his triangular figure, his squab and square form, is wanting in harmony throughout his entire person. The Malays seem to have sprung from a mixture of the Mongolian with White race, which often improves the type. The Papoo of New Guinea, in spite of the blackness of his skin, still preserves some advantages of form; but the South Australian, with his gaunt body, his lean members, his bending knees, his hump back, his projecting jaws, presents the most melancholy assemblage the human figure can offer. These portraits of an Australian warrior and of a native woman of Van Diemen's Land, show the last degree to which ugliness can go in this being, created so perfect, and destined to be the lord of all the world.

In the third direction, that of America, the same law makes itself felt. This face of an Oto Indian chief would have still some advantages, if the prominence of the cheek-bones, a slight elevation of the outer angle of the eyes, and the size of the jaw, did not clearly betray a less perfect nature. In the South American Indian, all these defects are still more exaggerated, and give to the races of the South, compared with those of the North, a very marked character of inferiority Finally, at the extreme point of the continent, and in

Terra del Fuego, live the Pecherays, the most misshapen the farthest from any culture, the most wretched, of all the inhabitants of the New World.

It would be still the same in advancing towards the poles. Passing the Finns, we arrive at the Laplanders: through the Mongolians, we reach the Tongoos, the Samoiedes of Siberia, and the Esquimaux of North America.

Thus, in all directions, in proportion as we remove from the geographical seat of the most beautiful human type, the degeneration becomes greater, the debasement of the form more complete. Does not this surprising coincidence seem to designate those Caucasian regions as the cradle of man, the point of departure for the tribes of the earth?

It results from this remarkable distribution of the races of man, that the continents of the North, forming the central mass of the lands, are inhabited by the finest races, and present the most perfect types; while the continents of the South, forming the extreme and far-sundered points of the lands, are exclusively occupied by the inferior races, and the most imperfect representatives of human nature. This contrast is more decided in the Old World than in the New; nevertheless, in the latter, notwithstanding the inferiority of the copper-colored race, we have seen that the man of the Northern race, the Indian of Missouri, has a marked superiority over the Indian of the South, over the Botocudes, the Guaranis, and the Pecherays of South America.

The degree of culture of the nations bears a proportion to the nobleness of their race. The races of the northern continents of the Old World alone are civilized; the southern continents have remained savage. In America, the civilized Aztecs of Mexico have come from the North. The ancient civilization of the Quichuas, at the summits of the Andes of Peru, scarcely seems itself indigenous to South America. It belongs elsewhere by its elevated position; it belongs to the temperate zone.

Now these differences between the North and the South are not of yesterday, nor to-day. If we consult the memorials of these tribes, without written history,—bounded and scanty as they are,—it might seem that it has been the same from a time ascending beyond all our traditions, if we except the Bible. No indication brings to light in these tropical continents the existence, at another epoch, of a purer type, of a more perfect race of men, than the inferior form we there meet with at the present day. The annals of the tribes in no part of these continents, record either the birth or the progress of a civilization which has contributed to the brilliant development of the present condition of man. Man has there always remained at the bottom of the scale of culture; while, from the earliest days of the world, history marks out the temperate continents as the seat of the refined communities. As there is a temperate hemisphere and a tropical hemisphere, we may, in the same manner, say there is a civilized hemisphere, and a savage hemisphere.

The distribution of man over the surface of the globe, and that of the other organized beings, are not then founded on the same principle. There is a particular law which presides over the distribution of the human races and of civilized communities, taken at their cradle, in their infancy; a different law from that which governs the distribution of plants and animals.

In the latter, the degree of perfection of the types is proportional to the intensity of heat, and of the other agents stimulating the display of material life. The law is of a physical order.

In man, the degree of perfection of the types is in proportion to the degree of intellectual and moral improvement. The law is of a moral order.

Thus the geographical march of the perfection of the species, from the poles to the equator, is suddenly broken when man appears, to recommence on a plan wholly new.

This difference between the two laws has its principle in the profound difference existing between the nature and destination of these distinct beings. The plant and the animal are not required to become a different thing from what they already are at the moment of their birth. Their *idea*, as the philosophers would say, is realized in its fulness by the fact alone of their material appearance, and of their physical organization. The end of their existence is attained, for they are only of a physical nature. But with man it is quite otherwise. Man, created in the image of God, is of a free and moral nature. The physical man, however admirable may be his organization, is not the true man; he is not an aim, but a means; he is not an end, like the animal, but a beginning. There is another, new-born, but destined to grow up in him, and to unfold the moral and religious nature until he attain the perfect stature of his master and pattern, who is Christ. It is the intellectual and the moral man, the spiritual man.

The law of development, if I may say so, is the law of man, the law of the human race, and of human societies; now, the free and moral being cannot unfold his nature without education; he cannot grow to maturity, except by the exercise of the faculties he has received as his inheritance.

Here is the reason, gentlemen, that the Creator has placed the cradle of mankind in the midst of the continents of the North, so well made, by their forms, by their structure, by their climate, as we shall soon see, to stimulate and hasten individual development and that of human societies; and not at the centre of the tropical regions, whose balmy, but enervating and treacherous, atmosphere would perhaps have lulled him to sleep the sleep of death in his very cradle.

Have we not the sad picture of what might have become of man, if he had had for his birth-place only the warm regions of the earth, in the wretched condition in which our unfortunate brethren of the inferior races still live, wandering to the furthest extremities of the tropical climates?

The fact of the gradual modification of the human types as we depart from a central race, seems to me to establish between all the varieties of mankind, however remote they may otherwise appear from the most perfect type, a bond of union, which, after having been established, science is not at liberty to pass over in silence, without taking into account. Now, if we consider the question of the formation of the races in the point of view we have just assumed, perhaps we shal see this field, once so dark, illuminated by some gleams of light.

However, before proceeding further, let us set forth one fact more, no less undeniable; for, in speaking of man, we must not forget there are always two sides to consider; the one physical, the other moral.

Western Asia is not only the geographical centre of the human race, but it is, moreover, the spiritual centre; it is the cradle of man's moral nature. Was it not there that those divine teachings were proclaimed, which the most cultivated communities in the world regard as their dearest treasure, and every man who loves the true, acknowledges to be Truth itself? Was it not there that the chosen people lived, to whom they were given in trust to preserve for the world until the time appointed by the Supreme Wisdom? Was it not there that the Saviour of all the members of the human family appeared, and the gospel of grace and liberty was preached, in the lowly valleys of Judea—that gospel which recognized neither Jew, nor Greek, nor Gentile, nor barbarian, and which invites all the races of the earth to salvation, without distinction? Is it not from the height of the sacred mount where He died upon the cross for all, that Christ bids every human soul, whatever be the ephemeral form of its earthly covering, to a spiritual union which he will consummate in his glory? And these great facts, gentlemen, interesting to every human being, these facts whose blessed consequences surround us on all sides at the present day, belong not to the number of those that any historical unbelief can ever strike out of the annals of mankind. Nor are they of secondary importance, considered merely in the results already accomplished; for who will maintain that, even in the future, man will ever witness an event more important for him than the appearance of the Saviour, and the proclamation of the universal gospel, destined to unite all men, and at the same time to bind them all to their common Creator?

Now, if man came from the hands of the divine Author of his being, pure and noble, it was in those privileged countries where God placed his cradle, in the focus of spiritual light, that he had the best chance to keep himself such. But how has he fallen elsewhere so low? It is because he was free, of a perfectible nature, and consequently capable also of falling. In the path of development, not to advance is to go back; it is impossible to remain stationary. The animal does not degenerate, because the form of his existence is necessary; he is not required to add anything. But man, who should grow in perfection by the constant exercise of the higher faculties of his nature, by struggling against the evil inclinations of a perverted will, man descends evermore, and proceeds from fall to fall, if he neglects those divine gifts, and abandons himself to the low instincts of his animal nature. He goes down to the life of the brute, whose form and semblance he takes. And what will come to pass if, separated from his God, and forgetting Him, he voluntarily stops the sources of the higher life, and moral life? Remote from the focus of tradition. where he might renew the temper of his faith, no remains unarmed in combat with that mighty nature that subjugates him; he yields in the struggle, and, vanquished, bears soon upon his figure the ineffaceable mark of bondage. Thus, perhaps, might one, I do not say explain, but conceive, the incontestable influence of each continent, and each region of the earth, on the physical forms, the character and the temperament of the man who dwells in it, and the degeneracy of his type in proportion as he is removed from the place of his origin, and the focus of his religious traditions. Renouncing moral liberty, which exists only in goodness, man gives to nature power over himself, submits to it, and thus are traced and distinguished, a race of Eastern Asia, an African race, an Australian race, a Polynesian race, an American race. We must confess, however, it is not granted to follow out, either in nature or in history, the steps of this transformation; a transformation that could only have taken place, at the time when the human race in their infancy had still the flexible and plastic nature of the child; and we must repeat, the origin of the human races is a fact beyond our observation and anterior to all history, and, like all other origins, is screened by an impenetrable veil.

Since man is made to acquire the full possession and mastery of his faculties by toil, and by the exercise of all his energies, no climate could so well minister to his progress in this work as the climate of the temperate continents. It is easy to understand this.

An excessive heat enfeebles man; it invites to repose and inaction. In the tropical regions the power of life in nature is carried to its highest degree; thus with the tropical man, the life of the body overmasters that of the soul; the physical instincts of our nature, those of the higher faculties; passion, sentiment, imagination, predominate over intellect and reason; the passive faculties over the active faculties. A nature too rich, too prodigal of her gifts, does not compel man to snatch from her his daily bread by his daily toil. A regular climate, the absence of a dormant season, render forethought of little use to him. Nothing invites him to that struggle of intelligence against nature, which raises the forces of man to so high a pitch, but which would seem here to be hopeless. Thus he never dreams of resisting this all-powerful physical nature; he is conquered by her; he submits to the yoke, and becomes again the animal man, in proportion as he abandons himself to these influences, forgetful of his high moral destination.

In the temperate climates all is activity, movement. The alternations of heat and cold, the changes of the seasons, a fresher and more bracing air, incite man to a constant struggle, to forethought, to the vigorous employment of all his faculties. A more economical nature yields nothing, except to the sweat of his brow; every gift on her part is a recompense for effort on his. Less mighty, less gigantesque, even while challenging man to the conflict, she leaves him the hope of victory; and if she does not show herself prodigal, she grants to his active and intelligent labor more than his necessities require; she allows him ease and leisure, which give him scope to cultivate all the lofty faculties of his higher nature. Here, physical nature is not a tyrant, but a useful helper; the active faculties, the understanding and the reason, rule over the instincts and the passive faculties; the soul over the body; man over nature.

In the frozen regions man also contends with nature, but with a niggardly and severe nature; it is a desperate struggle, a struggle for life and death. With difficulty, by force of toil, he succeeds in providing a miserable support, which saves him from dying of hunger and hardship during the tedious winters of that climate. No higher culture is possible under such unfavorable conditions.

The man of the tropical regions is the son of a wealthy house. In the midst of the surrounding abundance, labor too often seems to him useless; to abandon himself to his inclinations is a more easy and agreeable pastime. A slave of his passions, an unfaithful servant, he leaves his faculties, the talent God has endowed him with, uncultivated and unused. The work of improvement with him is a failure.

The man of the polar regions is the beggar, overwhelmed with suffering, who, too happy if he but gain his daily bread, has no leisure to think of anything more exalted.

The man of the temperate regions, finally, is the man born in ease, in the *golden mean*, the most favored of all conditions. Invited to labor by everything around him, he soon finds, in the exercise of all his faculties, at once progress and well-being.

Thus, if the tropical continents have the wealth of

nature, the temperate continents are the most perfectly organized for the development of man. They are opposed to each other, as the body and the soul, as the inferior races and the superior races, as savage man and civilized man, as nature and history. This contrast, so marked, cannot remain an open one; it must be resolved. The history of the development of human societies will give us the solution, or at least will permit us to obtain a glimpse of the truth.

LECTURE XI

The continent of the North considered as the theatre of history — Asia-Europe; contrast of the North and South; its influence in history; conflict of the barbarous nations of the North with the civilized nations of the South — Contrast of the East and West — Eastern Asia a continent by itself and complete; its nature; the Mongolian race belongs peculiarly to it; character of its civilization — Superiority of the Hindoo civilization; reason why these nations have remained stationary — Western Asia and Europe; the country of the truly historical races — Western Asia; physical description; its historical character; Europe — the best organized for the development of man and of societies; America — future to which it is destined by its physical nature.

LADIES AND GENTLEMEN: —

The result of the comparison we have made between the northern continents and the southern continents, in their most general characteristics, has convinced us, if I do not deceive myself, that what distinguishes the former is, not the wealth of nature and the abundance of physical life, but the aptitude which their structure, their situation, and their climate, give them, to minister to the development of man, and to become thus the seat of a life much superior to that of nature. The three continents of the North, with their more perfect races, their civilized people, have appeared as the *historical* continents, which form a marked contrast to those of the South, with their inferior races and their savage tribes.

Since this is the salient and distinguishing feature, securing to them decidedly the first place, we shall this evening proceed to study them more in detail as the theatre of history.

We know beforehand, gentlemen, that the condition of an active, complete development is the multiplicity of the contrasts, of the differences, — springs of action and reaction, of mutual exchanges exciting and manifesting life under a thousand diverse forms. To this principle corresponds, in the organization of the animal, the greater number of its special organs; in the continents, the variety of the plastic forms of the soil, the localization of the strongly characterized physical districts, the nature of which stamps upon the people inhabiting them a special seal, and makes them so many complicated but distinct individuals.

The various combinations of grouping, of situation, with regard to each other, placing them in a permanent relation of friendship or hostility, of sympathy or of antipathy, of peace or of war, of interchange of religions, of manners, of civilization, complete this work, and give that impulse, that progressive movement, which is the trait whereby the historical nations are recognized.

We may, then, expect to see the great facts of the life of the nations connect themselves essentially with these differences of soil and climate, with these contrasts, that nature herself presents in the interior of the continents, and whose influence on the social development of man, although variable according to the times, is no less evident in all the periods of his history.

Let us commence our inquiry with the true theatre of history — with Asia-Europe.

We have already had occasion to call attention to the unity of plan exhibited in this great triangular mass, which authorizes us to consider it as forming, in a natural point of view, a single continent, whose subdivisions bear the imprint of only secondary differences. We have also indicated, as the most remarkable trait of its structure, that great dorsal ridge, composed of systems of the loftiest mountains, traversing it from one end to the other in the direction of the length, which may even be regarded as the axis of the continent. It is, in fact, on the two sides of this long line of more than 5,000 miles, on the north and south of the Himalaya, of the Caucasus, of the Balkan, the Alps, and the Pyrenees, that the high lands of the interior of the continent extend. It splits Asia-Europe into two portions, unequal in size, and differing from each other in their configuration and their climate. On the south, the areas are less vast; the lands are more indented, more detached — on the whole, perhaps, more elevated; it is the maritime zone of peninsulas. On the north, the great plains prevail; the peninsulas are rare, or of slight importance, the ground less varied.

But what chiefly distinguishes one of the two parts from the other, what gives to each a peculiar nature, is the climate. Those lofty barriers we have just named almost everywhere separate the climates, as well as the areas. The gradual elevation of the terraces towards the south, up to this ridge of the continent, by prolonging in the southern direction the frosts of the north, augments still further, in Eastern Asia and in Europe, the difference of temperature between their sides, and renders it more sensible. Thus, almost everywhere, the transition is abrupt, the two natures wide apart. These high ridges arrest at once the icy winds of the poles, and the softened breezes of the south, and separate their domains. The Italian of our days, like the Roman of former times, boasts of his blue sky and his mild climate, and speaks with an ill-concealed contempt of the frosts and the ice of the countries beyond the Alps.

To the father of the Grecian poets, to Homer, who knows only the Ionian sky, the countries beyond the Hæmus are the regions of darkness, where rugged Boreas reigns supreme. At the northern foot of the Caucasus, the dry steppes of the Manytsch are swept by the frozen winds of the north; on the south, the warm and fertile plains of Georgia and of Imereth, feel no longer their assaults. In Eastern Asia, finally, the contrast is pushed to an extreme. The traveller, crossing the lofty chain of the Himalaya, passes suddenly from the polar climate of the high table lands of Tubet, to the tropical heats and the rich nature of the plains of the Indus and the Ganges. Yet, as we have said, this great wall, which separates the North from the South, is rent at several points. Between the Hindo-Khu and the Caucasus, the depressed edge of the table land of Khorasan, between the Caucasus and the Balkan, the plains of the Black Sea and of the Danube, open wide their gates to the winds and to the nations of the shores of the Caspian and the Volga. Between the Pyrenees and the Alps, the climates and the peoples of the South penetrate into the North.

Thus two opposite regions are confronted, one on the North, in the cool temperate zone, with its vast steppes and desert table-lands, its rigorous climates, its intense colds, its dry and starveling nature; the other on the South, in the warm temperate zone, with its beautiful peninsulas, its fertile plains, its blue heavens and its soft climate, its delicate fruits, its trees always green, its lovely and smiling nature.

The contrast of these two natures cannot fail to have a great influence on the peoples of the two regions. It is repeated, from the history of the very earliest ages, in the most remarkable manner. In the North the arid table lands, the steppes, and the forests, condemn man to the life of shepherds and hunters; the peoples are nomadic and barbarous. In the South, the fruitful plains and a more facile nature invite the peoples to agriculture; they form fixed establishments and become civilized. Thus in the very interior of the historical continent we find a civilized and a barbarous world, placed side by side.

Two worlds so different cannot remain in contact without reacting upon each other. The conflict begins, one might say, with history itself, and continues throughout its entire duration; there is scarcely one of the great evolutions, particularly in Asia, not connected with this incessant action and reaction of the North upon the South, and of the South upon the North, of the barbarian world upon the civilized world. At all periods we see torrents of barbarous nations of the North issuing from their borders and flooding the regions of civilization with their destroying waves. Like the boisterous and icy winds of the regions they inhabit, they come suddenly as the tempest, and overturn everything in their way; nothing resists their rage. But as after the storm nature assumes a new strength, so the civilized nations, enervated by too long prosperity, are restored to life and youth by the mixture of these rough but vigorous children of the North. Such is the spectacle presented by the history of the great monarchies of Asia and of their dynasties; that of Europe is scarcely less fertile in struggles of this kind. Some examples, which I proceed to recall to your memory, will be enough to convince you of the powerful influence of this contrast.

As far as the memorials of history ascend, it exhibits, on the table land of Iran and in the neighboring plains of Bactriana, one of the earliest civilized nations, the ancient people of Zend. The Zendavesta, the sacred book of their legislator, displays everywhere deep traces of the conflict of Iran, of the southern region, of the light of civilization — *the Good* — with the Turan, the countries of the North, the darkness, the barbarous peoples — *the Evil.* Who can say that even the idea of this dualism — of Good and Evil — which is the very foundation of the religious philosophy of Zoroaster, is not, to a certain extent, the result of the hostile relations between two countries so completely different? Six centuries before Christ, the barbarous Scythians come down from the North, sweep like a whirlwind through the same gate of the Khorasan upon the plateau of Iran, overrun the flourishing kingdom of Media, and spread themselves as far as Egypt. A whole generation was necessary to restore to Cyaxares his crown, and to efface the traces of this rude attack. In the eleventh century of our era, the Seldjouks Turks, descend from the heights of Bolor and Turkestan, invade first Eastern Persia, overturn the power of the Gaznevide Sultans, put an end to that of the Caliphs, and lord it over Western Asia. But nothing equals the tremendous shock caused through the whole of Asia by the invasion of the Mongolians. Issuing from their steppes and their deserts, under the conduct of the daring Gengis-Khan, the hero of his nation, their ferocious hordes extend like a devastating torrent from one end of Asia to another. Nothing withstands their onset; even Europe itself is threatened by these barbarians; all Russia is subjected, and scarcely can the assembled warriors of Germany drive them back from their frontiers, and save the nascent civilization of the West. China herself beholds a succession of conquerors establish in the North a brilliant empire, and for the first time the two Asias are subject to one and the same dominant people. India alone had been spared; she yields before a fresh invasion, and Sultan Babur — who already is no more a barbarian — founds, at the beginning of the sixteenth century, the mighty Mongolian Empire, which, in spite of its vicissitudes, has existed down to our days, and has yielded only to the power of the nations of civilized Europe. The history of China, lastly, is crowded with the struggles of the civilized people of the plain with the roving tribes of the neighboring table lands, and the last of these invasions, so frequent, — that of the Manchou Tartars, — has given to China its present rulers.

In Europe, the war of the North against the South,

though seemingly not so long continued, is not less serious. Six centuries before our era, bands of Celts, enticed by the attractions of the fertile countries of the South, set forth from Gaul, under the lead of Bellovese and Sigovese, cross the Alps, and proceed to establish themselves in the smiling plains of the Po. Other bands follow them thither, and found a new Gaul beyond the Alps. These impetuous children of the North soon press upon Etruria; and Rome, having drawn upon herself their anger, suffers the penalty of her rashness. About 390 B. C., the city was burnt, and the future mistress of the world well nigh perished in her cradle, by the strong hand of the very men of the North whom she was destined afterwards to subject to her laws. A century later, these same Gauls, finding Rome victorious and Italy shut against them, rush upon enervated Greece, give her up to pillage, and, profaning the sacred temple at Delphi, announce the fall of Hellas, and the last days of her glory and her liberty. Another troop of these bold adventurers cut their way into Asia Minor; they maintain themselves there, objects of terror in the land that bears their name, to the very moment when the power of Rome forced all the nations to bow beneath her iron yoke.

A century before the birth of our Saviour, the men of the North are again in motion. The Cimbri and the Teutons appear at the gates of Italy, and spread terror even to Rome herself. Forty years have scarce rolled away when Rome, in her turn, assails the Northern World. Cæsar marches to conquer the Gauls, formerly so terrible, and in the course of the ages, they are won to civilization. Thus, by the third gate which opens the wall of separation, the Southern World penetrates into that of the North.

But a still more earnest struggle then commences. The Germans have preserved their native energy, and are still free. Rome is declining, and, little by little, the sources of life in that immense body are drying up. The weaker it grows, the more the men of the North press upon the mighty colossus, whose head is still of iron, though its feet are of clay. It falls, for its own happiness and that of humanity; a new sap — the fresh vitality of the Northmen — is to circulate through it; and soon shall it be born again, full of strength and life.

You see, gentlemen, from the beginning to the end of history, the contrast of these two natures exercises its mighty influence. The struggle between the peoples of the two worlds is constant. In Asia it may be again renewed, for nature there is unconquerable, and the contrast still exists. In Europe, the coarse struggle of brute strength of the early days has ended, since, culture having passed into the North, conquerors and conquered, civilized men and barbarians, have melted down into one and the same people, to rise to a civilization far superior to the preceding. But we behold it reappear, less material but not less evident, between the free and intelligent thinker, and the Protestant of the North, and the artistic, impassioned, superstitious, Catholic man of the South.

Let us pass now to a second feature of the structure of the continent Asia-Europe, which has almost as much weight as that we have just discussed.

Long chains extending from the north to the south, in the direction of the meridians, the Bolor and Mt. Soliman, cut at right angles the great east-west axis. The Bolor forms the western margin of the high central plateau; the Soliman, the eastern margin of the table land of Iran; — the one on the north, the other on the south; — so that these two solid masses touch each other at their opposite angles, south-west and north-east. The remarkable point where these high ranges intersect, and the table land and the plains, lying outspread at their feet, touch each other, is the Hindo-Khu. These features of relief sever the continent into two parts, of almost equal extent, but of very unequal importance; Eastern Asia on the one side, and Western Asia and Europe on the other — the Mongolian races, and the White races.

This separation is so deeply marked in nature and in the nations, that even the ancients, with the practical sense belonging to them, made a division of Asia *intra Imaum*, and Asia *extra Imaum;* that is, Asia this side and Asia beyond the Bolor and the Hindo-Khu; as they also divided the North and the South into Scythia — Nomadic Asia — and Asia proper, or civilized Asia.

Eastern Asia forms, in fact, a continent by itself alone. A vast pile of highlands, a plateau in the form of a trapezium, occupies the entire centre, and forms the principal mass. It seems to invade everything; it is the prominent feature, and gives a distinctive physiognomy to the continent. It is surrounded on all sides by lofty ranges capped with snow, which seem, like towering ramparts, to guard it from attack, and to isolate it on every side. On the south the Himalaya, on the west the Bolor, on the north the Altai, on the east the Khingan, and the Yun-Ling, form an almost unbroken enclosure, whose detached summits belong to the loftiest mountains of the earth. A small number of natural entrances lead to the interior, or give an exit from it. The only gate that offers some facility is Zungary, between the Thian-Shan and the Altai; everywhere else, high and frozen passes.

The interior of this vast enclosure is cut by numerous chains, the highest of which—those of the Kuenlun on the south, and of the Thian-Shan on the north—are parallel to the Himalaya and the Altai, and divide the ground into several basins, or high bottoms. In all this extent, no fertile and easily cultivated plain; everywhere stretch the steppes, a dry and cold desert, or seas of drifting sand. Nevertheless, a considerable depression in Eastern Turkestan, where the Tarim flows, and whose bottom is marked by Lake Lop, allows the cultivation of the vine and the cotton tree at the foot of the Thian-Shan; but this is an exception. Apart from some privileged localities, nature here permits no regular tillage, and dooms the tribes of these regions to the life of shepherds and herdsmen,—the nomadic life.

Around this central mass, towards the four winds of heaven, extend at its feet broad and low plains, watered by the rivers pouring down from its heights, which rank among the largest in the world. On the north is the most extensive but the least important, the frozen and barren plain of Siberia, with the streams of the

Obi, the Jenisey, the Lena; on the east the low country of China, where meet and unite the two giant rivers of the Old World—those two twin rivers, which, born in the same cradle, flow on to die in the same ocean. On the south, the plain of Hindoostan, moistened by the fresh and abundant waters of the Himalaya, and the sacred streams of the Indus and the Ganges; on the west, finally, the plain of Turan, with the two rivers of Gihon and Sihon, and its salt seas, to which Western Asia already lays claim. In these plains, with fruitful alluvial soil, and on the banks of these blessed rivers, were developed the earliest, almost the only, civilized nations belonging to this continent. But the warm and maritime region of the East and the South, connected with the rich peninsulas of India, is by far the most favored of all. China and India, therefore, have given birth to the two great cultivated nations of Eastern Asia.

Nevertheless, as the great central ridge swerves obliquely towards the south, this warm and fortunate region forms only a narrow strip, not to be compared in extent with the cold, and sterile, and barbarous world of the North. This predominates, and decides its character.

Such are the distinctive features of Eastern Asia. What strikes us, in this world of the remotest East, is its gigantic proportions. The loftiest mountains of the earth, the most massive table lands, the most extensive plains, peninsulas which are small continents, rivers that have no rivals in the Old World, give it a character of grandeu and majesty not elsewhere to be found.

But, it is easily understood, nowhere are the differences also so strongly drawn, so huge, so invincible. Nowhere is the contrast between the high lands and the low lands, between the heat and the cold, between the moisture and the dryness, abundance and sterility, presented on so vast a scale. See, by the side of the low, burning, and productive plains of Hindoostan, ten or fifteen thousand feet higher up, the cold and arid highland plain of Tubet and Tangout; by the side of China and its populous cities, the elevated deserts and the tents of the nomads of Mongolia. The differences are everywhere pushed to their utmost limit.

Furthermore—and this characteristic completes the picture—the communications from one region to another are always difficult. One thoroughfare alone, the valley of the Peschawer, leads from Persia to India, and has been the highway of all the conquerors, from Alexander to Babur and to the English. No practicable road for armies or for regular commerce unites India and China: the peninsulas communicate only by sea. The passes of the Himalaya are at an elevation of from ten to eighteen thousand feet; those of the Bolor are frozen in the middle of summer. At all times, the passage of the plateau is a difficult and tedious undertaking, and at certain points almost impossible.

Eastern Asia is, then, preëminently, the country of contrasts, of isolated and strongly characterized regions; for each forms a world apart, and is sufficient unto itself.

What must be the effect of this strong and massive

nature upon the nations who live under its influence history will inform us.

As Eastern Asia has a physical nature belonging especially to itself, so it has a particular race of men— the Mongolian race. We have already pointed out the external characteristics of the Mongolian family. With it, the melancholic temperament seems to prevail; the intellect, moderate in range, exercises itself upon the details, but never rises to the general ideas or high speculations of science and philosophy. Ingenious, inventive, full of sagacity for the useful arts and the conveniences of life, the Mongolian, nevertheless, is incompetent to generalize their application. Wholly turned to the things of earth, the world of ideas, the spiritual world, seems closed against him. His entire philosophy and religion are reduced to a code of social morals limited to the expression of those principles of human conscience, without the observance of which society is impossible.

The principal seat of the Mongolian race is the central table land of Asia. The roaming life and the patriarchal form of their societies are the necessary consequence of the sterile and arid nature of the regions they inhabit. In this social state, the relations and the ties which unite the individuals of the same nation are imposed by kindred, by birth; that is, by nature. Association is compulsory, not of free consent, as in more improved societies. Thus, the greater part of Eastern Asia seems doomed to remain in this inferior state of culture; for the whole North—Siberia and its vast areas—is scarcely more suited to favor the unfolding of a superior nature.

Nevertheless, in the warm and maritime zone, in the fertile and happy plains of China and India, along those rivers which support life and abundance on their banks, nations, invited by so many advantages, establish themselves and fix their dwelling-places. Their number soon augments; they demand their support from the soil, which an easy tillage yields them in abundance. They become husbandmen; cultivated societies are formed; civilization rises to a height unknown to the tribes of the table-land.

The Chinese, of Mongolian race, preserves, even in his civilization, the character as well as the social principle stamped upon his race by nature—the patriarchal form. The whole nation is a large family; the Emperor is the father of the family, whose absolute, despotic, but benevolent power governs all things by his will alone. China, then, in the order of civilized nations, is the purest representative of Eastern Asia, and shows to what point the patriarchal principle of the earliest communities is compatible with a higher cultivation.

In India, the nations of the white race, sprung from the West, have founded a civilization wholly different, the character of which is explained at once by the primitive qualities of the race and the climate.

Endowed with a higher intelligence, with a power of generalization, with a profound religious sentiment, the Hindoo is the opposite of the Chinese; for him the invisible world, unknown to the Chinese, alone seems to exist. But the influence of the climate of the tropics gives to the intuitive faculties an exaggerated preponderance over the active faculties. The real, positive world disappears from his eyes. Thus, in his literature, so rich in works of high philosophy, of poetry and religion, we seek in vain for the annals of his history, or any treatise on science, any of those collections of observations so numerous among the Chinese. In spite of these defects the Hindoo civilization, compared to that of China, bears a character of superiority, which betrays its noble origin. It is the civilization of the western races transported and placed under the influence of the East.

But there is one characteristic common to all these civilizations of the uttermost East, deserving our particular attention. Born in the earliest ages of the world, (for without admitting—far from it—the fabulous antiquity their own traditions assign them, we may regard them as belonging to the most ancient in the world,) they seem to grow rapidly at first, and at the remotest period recorded by history they have already acquired the degree of development and all the leading features that distinguish them at the present day. Nearly fifteen hundred years before Christ—others say two thousand—India already possessed the Vedas—those religious and philosophical works, which already suppose a high culture and its accompanying social state. Alexander finds it flourishing and brilliant still, but little changed; the description the historians of his conquests have left, is true of modern India when invaded by the English. As much may be said of China, whose existing condition seems to present the same essential features we know it to have possessed from a time long before our era. Thus, these nations offer the astonishing spectacle of civilized communities remaining perfectly stationary. Three thousand years of existence have made no essential change in their condition, have taught them nothing, have brought about no real progress, have developed none of those great ideas, which effect a complete transformation in the life of nations. They are, as it were, stereotyped.

What, then, has been wanting to these peoples, that they have not been favored with a further progress? Why do they all stop short in the career they have entered upon in so brilliant a manner — even the Hindoos of noble race — of the race eminently progressive?

What has been wanting to the communities of Eastern Asia, ~~gentlemen,~~ is the possibility of actions and reäctions upon each other, more intimate, more permanent; it is the possibility of a common life.

These nations are too isolated by nature, too opposite in race and character, to be able to blend in one common civilization. The Hindoos are separated from China by the snowy terraces of the Himalaya, and of the Yun-Nan; from Western Asia, by the high table lands of Caboul. These forms of relief are too huge; the contrasts resulting from them are too violent; they are unconquerable by man. Meantime, each of these rich districts may suffice, of itself alone, for a beautiful career of improvement, but each of these peoples furnishes but a single type. In their isolation, their excellences, as well as their defects, run into excess; nothing tempers or corrects them; their character is more individual. Such is the strength of these civilizations, that clouds of conquerors are successively absorbed, without modifying them, almost without leaving a trace behind.

~~But individuality is here carried to egoism.~~ Of this very isolation which causes their inferiority, and kills all progress, they make a conservative principle. The Hindoo cannot leave his country except by sea; the Vedas forbid it under pain of pollution. Japan and China obstinately close their borders against all the nations of Europe, and it is only at the cannon's mouth that the English have opened the gates so long shut, and forced them to the life of interchange which will restore them to progress and vitality. Thus, while everything around them is advancing, India and China have remained stationary. For it is not given to one people alone, any more than to one individual alone, to run through the whole compass of the scale of human progress, by themselves and without the aid of their brethren. Eastern Asia is, then, the continent of extreme contrasts and of isolated regions; of races essentially Mongolian; of stationary civilizations; of the semi-historical nations. It is not there that the work of the development of humanity can be achieved.

The second half of the Old World, in the temperate region, Western Asia and Europe, forms another whole, wherein we are able to point out several common characteristics. Besides the division into a North and a South, on the two sides of the continental axis, the most salient feature is the long table land of Iran, stretching uninterruptedly from India to the extremity of Asia

Minor, and even prolonging itself, without losing its nature, across the peninsulas of the Mediterranean, as far as Spain.

From one end of these regions to the other, nature wears a character of uniformity. Everywhere the same cretaceous and jurassic limestone deposits form the greater part of the ground; everywhere volcanoes rise from the earth, and shake it with their convulsions. The climate, also, is alike; for in Asia a more southern latitude is counterbalanced by a greater elevation of the plateaus. The flora is analogous; the cultivated plants, the fruits, the domestic animals, are the same, with the exception of the camel of the desert, useless to Europe. Finally, the white Caucasian race, the most noble, the most intellectual of the human species, and all the nations of progressive civilization, dwell there. If we add Egypt and the vicinage of the Atlas, which belong to the Mediterranean, it is the true theatre of history, in the proper meaning of that word.

Nevertheless, in spite of this real community of characteristics, it is easy to detect, in Western Asia and Europe, certain differences not less important, that force us to consider them still as two distinct continents.

In Europe, in the southern zone, the plateau loses its continuity, and splits into peninsulas. In the northern zone, the arid steppes and the deserts are changed beyond the Oural into a fertile soil, more elevated, well watered, covered with forests, and susceptible of cultivation. The areas become gradually smaller, and the whole continent is only a great peninsula, of which the headland turning towards the west juts out into the ocean. The north-east direction of the continental axis crowding the lands further north, and the influence of the ocean, give it a wetter and more temperate climate. Let us further examine these two portions of Asia-Europe, considered in the historical point of view.

Western Asia is placed in the middle portion of the continent Asia-Europe, between the two extremes. Like Eastern Asia, it has for its centre and prominent feature, a table land encircled with mountains, the plateau of Iran and of Asia Minor; but it is narrower, more elongated. The mountain chains are less elevated, less continuous. The mountains of Kurdistan and of the Taurus, which edge it on the south, attain a height of ten or twelve thousand feet only at a few points. The higher mountains, as the Ararat, are isolated, or form a chain, detached from the mass, like the Caucasus. We have already said that the north-east side is low and entirely open. The deep valley of Peschawer cuts its eastern side, and opens a passage towards India. Not only is this plateau more accessible than that of Eastern Asia, by reason of these forms of relief, but very different from the latter, which is far from any ocean, it is bathed at its very feet, on the four corners, by inland seas, that are so many new outlets. On the south, the Arabian Sea, the Persian Gulf, and the Mediterranean; on the north, the Caspian and the Black Sea.

Low and fertile plains, watered by twin streams, stretch at the foot of the table land of Iran. On the south, the plain of the Euphrates and the Tigris, the unequalled fertility whereof ceases with the rich alluvial lands of those rivers; on the north, the no less happy plains of Bactriana, watered by the Gihon and the Sihon. Beyond these lifegiving rivers, the steppes of the deserts establish their empire.

The climate of Western Asia no longer offers those extreme contrasts which strike us in Eastern Asia. The plateau is on the south of the central ridge, and not on the north, and enjoys a favored climate. It is less dry, more fertile; the desert there is less continuous; these southern plains are not under the tropics; the difference between the plain and the table land is softened.

The true Western Asia, the Asia of history, is reduced thus to a plateau flanked by two plains. Add the Soristan, which connects it with Egypt and this last-mentioned country, and you will have all the great countries of civilization at the centre of this continent: on the north, the nomads of the steppes of the Caspian; on the south, the nomads of Arabia and its deserts form the natural limits of the civilized world of these countries. Compared with the East, the areas are less vast, the reliefs less elevated, the nature less continental — notwithstanding its more central position — the contrasts less strongly pronounced, the whole more accessible.

Here, as we have said, is the original country of the white race, the most perfect in body and mind. If we take tradition for our guide, and follow step by step the march of the primitive nations, as we ascend to their point of departure, they irresistibly lead us to the very centre of this plateau. Now, in this central part also, in Upper Armenia and in Persia, if you remember, we find the purest type of the historical nations. Thence we behold them descend into the arable plains, and spread towards all the quarters of the horizon. The ancient people of Assyria and Babylonia pass down the Euphrates and the Tigris into the plains of the South, and there unfold, perhaps the most ancient of all human civilization. First, the Zend nation dwells along the Araxes, then, by the road of the plateau, proceeds to found, in the plains of the Oxus, one of the most remarkable and the most mysterious of the primitive communities of Asia. A branch of the same people, or a kindred people — the intimate connection of their language confirms it — comes down into India, and there puts forth that brilliant and flourishing civilization of the Brahmins, of which we have already spoken. Arabia and the North of Africa receive their inhabitants by Soristan; South Europe, perhaps, by the same routes, through Asia Minor; the North, finally, through the Caucasus, whence issue, in succession, the Celts, the Germans, and many other tribes, who hold in reserve their native vigor for the future destinies of this continent. *There* then is the cradle of the white race at least — of the historical people — if it is not that of all mankind.

The civilizations of Western Asia also, as well as those of Eastern Asia, spring up in the alluvial plains which are easily tilled, and alike connect themselves with the great rivers, and not, as in Europe, with the seas. The plains of Babylonia and of Bactriana are continental, and not maritime, like India and China. The contrasts of nature are still strongly expressed, but yet less so than in the East. There are still vast spaces, and consequently vast states. The religions, the political and social condition of the people, still betray the influence of a nature man has not yet succeeded in overmastering.

The civilizations are still local, and each has its special principle; and yet there is no more of isolation. The accessible nature of all these regions, as we have seen, makes contact easy, and facilitates their action upon each other; a blending is possible, and it takes place. The formation of great monarchies, embracing the whole of Western Asia, from India to Asia Minor, from the steppes of Turan to the deserts of Arabia, is a fact renewed at every period of their history. Assyria, Babylonia, Persia, reünite successively under the dominion of the same conqueror all these various nations. But no one knew so well as Alexander how to break down all the fences that kept them apart. The lofty idea which reigned in the mind of that great conqueror, that of fusing together the East and the West, carried with it the ruin of the special civilizations of the East and the universal communication of Hellenic culture, which should combine them in one spirit, and drew the whole of that part of the world into the progressive movement Greece herself had impressed on the countries of the West.

Egypt, alone, in her isolation, represents, up to a certain point, the nature of Eastern Asia. Yet she, too, was compelled to yield to the social and progressive spirit of Greece, which soon brought her into the circle of relations with the nations of the West.

Thus the people and the civilizations of Western Asia were saved from the isolation and egoism so fatal to China and to India. They perished in appearance, but it was only to sow among the very nations who were their conquerors, the prolific seeds of a fairer growth, whereof the future should gather the fruits.

Europe, in her turn, has a character quite special, whose principal features we have already pointed out. Although constructed upon the same fundamental plan with the two Asias, it is only the peninsular headland of all this continent. Here are no more of those gigantesque forms of Eastern Asia, no more of those boundless spaces, no more of those obstacles against which the forces of man are powerless, of those contrasts that sunder the opposite natures, even to incompatibility. The areas contract and shrink; the plateaus and the mountains are lowered; the continent opens on all sides. None of those mortal deserts to cross, none of those impassable mountain chains, which imprison the nations. From the foot of Italy to the head of Cape North, from the coasts of the Atlantic to the shores of the Caspian, there is no obstacle a little art may not overcome without much effort. The whole continent is more accessible; it seems more wieldy, better fashioned for man.

And yet, gentlemen, all the contrasts of both Asias exist, but they are softened, tempered. There is a Northern World, and a Southern World, but they are less different, less hostile; their climates are more alike. Instead of the tropical plains of India, we find there the fields of Lombardy; instead of the Himalaya, the Alps; instead of the plateaus of Tubet, those of Bavaria. The contrasts are even more varied, more numerous still.

The table land of the South is broken up into peninsulas and islands; Greece and its archipelago, Italy and its isles, Spain and its sierras, are so many new individuals, exciting each other reciprocally to animation. The ground is everywhere cut and crossed by chains of mountains, moulded in a thousand fashions, in such a way as to present, within the smallest possible space, the greatest number of districts physically independent.

Add to all these advantages that of a temperate climate, rather cold than hot, requiring of men more labor and effort, and you will be satisfied that nature is nowhere better suited to lift man by the exertion of his powers to the grandeur of his destination.

Nevertheless, the earliest civilized societies do not spring up in Europe; she is too far removed from the cradle of the nations, and the beginnings are less easy there. But these first difficulties once overcome, civilization grows and prospers with a vigor unknown to Asia. In Asia it is in the great plains, on the banks of the rivers, that civilization first shows itself. In Europe, it is on the peninsulas and the margin of the seas.

Europe is thus the most favored continent, considered with respect to the education of man, and the wise discipline it exercises upon him. More than any other it calls into full play his latent forces, which cannot appear and display themselves, except by their own activity. Nowhere can man better learn to subdue nature, and make her minister to his ends. No continent is more fitted, by the multiplicity of the physical regions it presents, to bring into being, and to raise up so many different nations and peoples.

But it is not alone for the individual education of each people that Europe excels; it is still more admirably adapted than any other continent to favor the common relations of the countries with each other, to increase their reciprocal influence, to stimulate them to mutual intercourse. The smallness of the areas, the near neighborhood, the midland seas thick strown with islands, the *permeability* of the entire continent —pardon me the word—everything conspires to establish between the European nations that community of life and of civilization which forms one of the most essential and precious characteristics of their social state.

America, finally, the third continent of the North, presents itself to us under an aspect entirely different. We are already acquainted with its structure, founded on a plan widely departing from that of Asia-Europe; we know that its characteristic is simplicity, unity. Add to this feature its vast extents, its fruitful plains, its numberless rivers, the prodigious facility of communication, nowhere impeded by serious obstacles, its oceanic position, finally, and we shall see that it is made, not to give birth and growth to a new civilization, but to receive one ready-made, and to furnish forth for man, whose education the Old World has completed the most magnificent theatre the scene most worthy of his activity. It is *here* that all the peoples of Europe may meet together, with room enough to move in, may commingle their efforts and their gifts, and carry out, upon a scale of grandeur hitherto unknown, the life-giving principle of modern times—the principle of free association.

The internal contrasts which assisted the development of the nations in their infancy and youth, exist not here; they would be useless. They are reduced to two general contrasts, which will preserve their importance; the sea-shore and midland on the one side, and the North and the South on the other. The last will be further softened down, when slavery, that fatal heritage of another age, which the Union stills drags after it, as the convict drags his chain and ball, shall have disappeared from this free soil, freed in the name of liberty and Christian brotherhood, as it has disappeared from the fundamental principles of its law.

Thus America also seems invited, by its physical nature and by its position, to play a part in the history of humanity, very different indeed from that of Asia and Europe, but not less glorious, not less useful to all mankind.

LECTURE XII.

Geographical march of history — Asia the cradle of civilization — Common character of the primitive nations — Powerful influence of nature — The human race in its infancy lives under authority, which becomes slavery — Civilization passes to Europe — Greece; period of youth; emancipation, and intellectual and moral development; action on the East and West; the Greek the teacher of the world — Rome; her work, political and social — Inability of the Ancient World to attain the end of humanity — Coming of Christ; his doctrines new in a historical point of view — The Germanic Christian world begins their application — Civilization passes to the North, and embraces all Europe; its different phases — Europe owes it to the rest of the world — Discovery of America — Universal inroad of the civilized nations — Social work begun at the same time — America must finish it — The people of the future; by what signs recognized — Conclusions — Foreseen solution of the contrast of the three Northern continents and the three Southern — Duties of the privileged races towards the inferior — A few words upon the method pursued — Science and faith.

Ladies and Gentlemen :—

The examination we have made of the structure of the northern continents, considered in respect of the influence they exercise through their physical nature upon the condition of human societies, enables us to judge in advance that they are formed to act different parts in the education of mankind. It remains to be seen whether the course of history will confirm these anticipations. Now, if we find a real concordance, a harmony between these two orders of facts, we may fearlessly assert that these differences of physical organi zation were intentional, and prepared for this end by Him who controls the destinies of the world.

The first glance we cast upon the annals of the nations, enables us to perceive a singular but incontestable fact, that the civilizations representing the highest degree of culture ever attained by man, at the different periods of his history, do not succeed each other in the same places, but pass from one country to another, from one continent to another, following a certain order. This order may be called *the geographical march of history.* We now proceed to set this forth by a rapid review of the great phases through which human societies have passed in their gradual improvement.

Asia, the country of the superior races, Western Asia, above all, the country of the white race, of the historical race, is also the cradle of the earliest civilized communities whose existence is commemorated by history.

Tradition universally represents the earliest men descending, it is true, from the high table lands of this continent; but it is in the low and fertile plains lying at their feet, with which we are already acquainted, that they unite themselves for the first time in national bodies, in tribes, with fixed habitations; devoting themselves to husbandry, building cities, cultivating the arts; in a word, forming well-regulated societies. The traditions of the Chinese place the first progenitors of that people on the high table land, whence the great rivers flow; they make them advance, station by station, as far as the shores of the ocean. The people of the

Brahmins come down from the regions of the Hindo-Khu, and from Cashmere, into the plains of the Indus and the Ganges; Assyria and Bactriana receive their inhabitants from the table lands of Armenia and Persia.

These alluvial plains, watered by their twin rivers, were better formed than all other countries of the globe, to render the first steps of man, an infant still, easy in the career of civilized life. A rich soil, on which overflowing rivers spread every year a fruitful loam, as in Egypt, and one where the plough is almost useless, so movable and so easily tilled is it, a warm climate, finally, secure to the inhabitants of these fortunate regions plentiful harvests in return for light labor. Nevertheless, the conflict with the river itself and with the desert, which, on the banks of the Euphrates, as o those of the Nile and the Indus, is ever threatening to invade the cultivated lands, the necessity of irrigation, the inconstancy of the seasons, keep forethought alive and give birth to the useful arts and to the sciences of observation. The abundance of resources, the absence of every obstacle, of all separation between the different parts of these vast plains, allow the aggregation of a great number of men upon one and the same space, and facilitate the formation of those mighty primitive states which amaze us by the grandeur of their proportions.

Each of them finds upon its own soil all that is necessary for a brilliant exhibition of its resources. We see those nations come rapidly forward and reach in the remotest antiquity a degree of culture, of which the temples and the monuments of Egypt and of India, and the recently discovered palaces of Nineveh, are living and glorious witnesses.

Great nations, then, are separately formed in each of these areas, circumscribed by nature within natural limits. Each has its religion, its social principles, its civilization severally. But nature, as we have seen, has separated them; little intercourse is established between them; the social principle on which they are founded is exhausted by the very formation of the social state they enjoy, and is never renewed. A common life is wanting to them; they do not reciprocally share with each other their riches. With them movement is stopped; everything becomes stable and tends to remain stationary.

Meantime, in spite of the peculiar seal impressed on each of these Oriental nations by the natural conditions in the midst of which they live, they have, nevertheless, some grand characteristics common to all, some family traits that betray the nature of the continent and the period of human progress to which they belong, making them known on the one side as Asiatic, and on the other side as primitive.

The causes of this phenomenon are at once of a moral nature and of a physical nature.

Man is still in the period of infancy, and infancy must needs be trained under the authority of a law which guides his first steps. Even by virtue of an inward nature, of a moral nature reflecting the divine image of his Maker, he cannot grow up to complete development, to his perfect stature, except by fulfilling the will of Him who calls him to such lofty destinies. This will is the supreme good; all that departs from it is evil. Man created free must fulfil it freely, and with consciousness of its excellence; but this very liberty, the most infallible sign of the nobleness of his nature, conceals the danger of a fearful fall. This liberty led the men of the earliest times on to that pitch of wickedness which rendered necessary the first great catastrophe of the human race, that earliest great punishment of the Flood, of which all nations, even the most barbarous, have preserved an appalling memory. Above all things it is the duty of man, if the work of his discipline is not to stop-short of its end, to learn his dependence upon the Judge of good and of evil; to learn that saving fear of God which is the beginning of wisdom, and which alone can regulate the employment of his liberty, and hinder him from surrendering himself to the irregular inclinations of his finite nature. Now, God had revealed himself to man; had made known to him his will, and pointed out the path which he ought to have followed. The Creator himself condescended to guide the steps of the creature upon the long journey he had to travel. This is what the Bible tells us; this is confirmed by the vague memorials of all the primitive nations, whose oldest traditions, those antecedent to the philosophical Theogonies prevalent at a later period, and giving them their specific character, contain always some disfigured fragment of this divine history.

But man soon became unfaithful; like a true prodigal son, he abandoned the benevolent Father, under whose protection he was living; he cast off the yoke easy to bear; he forgot the living God who had been revealed to him, and, submitting to the lower instincts of his being, he fell under the power of nature.

Recall, meantime, to your minds, gentlemen, all that we have learned of the stupendous and massive forms of that Oriental nature, of its insuperable contrasts, of its climate, tropical in India and in a part of China, very hot still on the banks of the Euphrates and the Nile; of that physical vigor which the Old World displays upon all the points favored by the copiousness of the waters, and you will understand that man, a child still, brought into the presence of such a power, must have felt himself, not merely a dependent, but a slave. The river he looks to for the fertility of the soil that feeds him — the animal, the plant, that minister to his wants — the sun, above all, that bright orb which reigns over nature, and in alternate march seems to dispense either life or death at his will — everything becomes to him an object of worship. He acknowledges the powers of nature as his gods, to whose mercy he feels himself to be committed, and accepts for his supreme rule the inflexible law that governs the heavenly bodies. He is falling from the world of liberty into that of necessity.

After this, what reason is there to be astonished that everything in those ancient civilizations bears the impress of the subjection of human liberty to the yoke of nature? All the religions, however varied they may otherwise appear, are the worship of the heavenly hosts. The immutable, blind laws of necessity, regulating the courses of the celestial bodies and the life of nature, these are the gods of the early East; inflexible, despotic, unloving, inexorable.

There all science appears as traditional. Man attains not to the light by his own activity. The truth is not the recompense of his efforts, of his progress, of the free unfolding of his faculties. It is transmitted to him already prepared from elsewhere.

In social life, casts, separated by insurmountable barriers consecrated by religion itself; or, in the patriarchal state, domestic relations, imposed by nature, restrain the free movement of the human faculties.

In political life, absolute monarchy, the entire organization of which is only the earthly image of the great Celestial Court of the Sun and his retinue, and of which the chief, representative of the Deity himself, is clothed with an unlimited power like him, and like him pronounces irrevocable decrees.

Such are the features common to all the civilized communities of the early East; one people alone forms an exception,—poor and insignificant in appearance, but great in its destinies;—it is the Jewish people, the people of God. In the midst of the defection of all the nations, they received the glorious mission of preserving in the world the knowledge of the only personal, living, and true God. Placed under His law, they would have been able to show, had they remained always faithful, what man might have become under the paternal government of His Creator; but their history is scarcely anything but that of disobedience and chastisement, and it enables us the better to see that at this first period of his development, man is under the law, and not under the economy of grace and liberty.

During the long centuries of these first ages, man has

therefore learned but one thing, that he depends on the will of a master, but that master is an inexorable despot, devoid of love. He can only fear him; if he obeys him, it is as a slave; he loves him not, nor adores him, for love presupposes liberty.

Man cannot, gentlemen, remain thus. A cry of liberty makes itself heard; it reëchoes to the depths of that East which groans in its chains; it issues from the land of the West, a land of emancipation and liberty; from that Europe which in a thousand various ways allures man to the free culture of his faculties. In a small corner of the earth, neighboring still to the East, but admirably organized, in that small peninsula of Greece, where all the varied contrasts of the whole continent seem to be repeated in a narrow space, under a climate blessed of Heaven, a new people arise, upon a new land, a free people, a people of brethren. With them the period of youth commences; human consciousness awakes with energy; man recovers himself; the slave bent beneath his yoke springs up and holds his head erect. The Greek, with his festivals, his songs, his poetry, seems to celebrate, in a perpetual hymn, the liberation of man from the mighty fetters of nature.

A new civilization is to be born; all these riches of poetry, of intellect, of reason, which are the heritage of the human mind, display themselves without obstacle, and expand in the sun of liberty. Who can describe all there is of fresh and youthful energy in that people of artists and philosophers, whose efforts open to us a world entirely new? This is no longer the world of nature; it is that of the human soul. Everything, in fact, with the

Greek bears that eminently human character which betrays the preponderance of human personality and the energy of individual character.

His religion is a deification of the faculties and affections of man. In place of the passionless, immovable deities of Egypt and of Persia, his Olympus presents the animated spectacle of an assembly of human persons, free and independent, presided over by the happy conqueror of the elder gods of nature. Destiny, banished almost beyond the confines of heaven, hardly reminds us of those blind and deaf gods, those gods of necessity. who reigned absolutely over all the East. When the forces of nature, when the trees of the forests, the mountains, the springs, and the rivers, appear as objects of worship, it is under the form of gods, of goddesses, and of nymphs, endowed with all the affections, and subject to all the weaknesses of common mortals.

Greek science is no longer merely traditional; we see its birth and its growth; it is the production of the efforts of the human soul; it is progressive; the Greek no longer goes to the outer world of nature in search of wisdom, but descends to the depths of human consciousness. With Socrates and his school, philosophy has passed from the realm of nature into the realm of man; she has become a moral philosophy.

In the social life of the Greeks, no more castes, no more of those hard sacerdotal despotisms of the East, which, by regulating human existence in detail, hinder its improvement; but communities of free and equal men and the predominance of democracy, that is, of individual and local life; these are its characteristics.

Such is the impulse the awakening of human personality impresses on this chosen people, that a few centuries suffice to achieve the work of the most brilliant display of the human mind, and of a culture leaving far behind all the nations of the East. Among them all the flowers of genius bloom together; their poets, their sculptors, their historians, their philosophers, have been, down to our day, and will hereafter be, the guides and the models of the man of taste and intelligence, in all countries and in all ages. The Greek becomes the teacher of the whole world.

The civilization of the Greeks is a conquest of man too beautiful to remain confined within the narrow limits of this petty country and inconsiderable people; all mankind must needs enjoy the benefit. The East, having given so much to Greece during her infancy, possessed the first rights in the achievements of her maturity. The conquests of Alexander begin the work of planting Grecian culture in the ancient soil of Asia, in the bosom of those worn-out nations which seem ready to perish in their weakness. A fresh sap flows through them, and Western Asia, drawn forcibly into the movement of the nations of the West, henceforth takes her part in their progress and their vicissitudes. Eastern Asia alone is untouched, and remains stationary. India and China, fossil remains of that ancient Orient which perished under the blows of the Greeks, subsist, as if to represent, down to the present moment, the antique civilization of the first ages, and to show the imbecility of its principle. At a later period, Rome, with her rude warriors, comes herself to seek for culture and the arts on the soil of Greece; and Greece, conquered by arms, still reigns by her genius over her very conquerors.

Nevertheless, gentlemen, the Greek, who carried the individual culture of man to so high a pitch, knew not how to establish the social relations on a solid basis, nor to organize a national body, nor to combine the peoples subjected to his influence into a system of nations strongly united together. I wish for no other proofs than that terrible Peloponnesian war, that fratricidal struggle, from which dates the decline of Greece, and the lamentable history of the Empire of Alexander and his successors. The Greek principle is individuality, and not association, and this is still further determined by the race, by the tribe; that is, by nature, and not by voluntary agreement.

This political and social work is a new work, and is entrusted to a new country and a new people. The centre of the civilized world again changes place; it takes a step further towards the West; its circumference enlarges; it embraces at once the South, the East, and the West. Rome, more skilled in the arts of conquest, and of establishing solid and durable political ties between the nations, combines in one and the same social net-work all the civilized nations of the Ancient World. The place she occupies in the very middle of the basin of the Mediterranean, seems to foretell that she is destined to become the metropolis of all the cultivated peoples who dwell upon its shores. This vast empire recombines the various elements of all the foregoing epochs in one and the same civilization, and the Roman world, having profited by all these advantages, offers the spectacle of the most brilliant social epoch of which the history of antiquity has anything to say.

And yet, in spite of all these advances, if we look somewhat nearer, what inability to accomplish the aim of humanity, what universal selfishness and corruption! No common faith binds together the nations, aggregated, rather than united. Rome exacts only one worship, that of the Emperor, who personifies the state. On all sides, conquerors and conquered still are found, and in this land of liberty one half of the men are slaves to the other. The Roman world, like all the rest, is to perish by its own vices.

Thus far, as you see, gentlemen, man has attempted to go his own way, growing up without God. He has not, however, been abandoned, as his progress shows; but he has exhausted all the spells and conjurations this procedure enabled him to try. He is convinced of his weakness; doubt takes hold of him and devours him; despair stands at his gate. All the literature of the Roman Empire confirms this. He has passed from the idolatry of nature to that of man; from the idolatry of man to that of society, represented by the head of the state. He must return to the true God, or there is no hope for him in the future.

It was then that the meek form of the Saviour appeared upon the scene of the world. What comes he to teach upon the earth? He recalls man to the only God, personal, free, full of love, merciful, the God of salvation. He proclaims the equal worth of every human soul, for he died for all. He gives unto men that new commandment, "Love one another as I have loved you," for ye are all brethren, and children of the same Father.

Thus, no more idolatry, no more servitude; for he liberated man from the yoke of evil that restrains the freedom of his moral being. No more thraldom; for that is incompatible with the rights of his brethren and with the love he owes them. No more national religions, opening between the nations abysses that nothing can fill up. All the nations of the earth must unite together in spirit, by the bonds of the same faith, under the law of the same God. This is the lofty goal to which henceforth all human societies ought to aim. The world hears the unity and brotherhood of all human kind proclaimed, without distinction of nation or of race — the true principle of humanity. This is the leaven that is to leaven the whole lump; it is upon this new basis that humanity, recommencing its task, goes on to build a new edifice.

But what people shall be charged with this immense work? Shall it be that old Roman society, wholly pagan still in its origin and in its forms, stained by slavery and violence, condemned long since to perish for its crimes? That body whose sap is gone, whose principle of life is exhausted, whose work is finished, — can it be born again? No, gentlemen, it is glory enough for the Roman world to have received and borne in its bosom this precious seed of the future, and to have shielded its earlier growth. The Church had her birth there, but the Christian world must needs bloom elsewhere.

The North is summoned in turn; the fierce Germans, after five centuries of struggle, break down the old empire, but adopt Christianity. In the midst of this great and universal ruin, the Church alone remains upright, and becomes the corner stone of the new edifice. Civilization passes to the other side of the Alps, where it establishes its centre. A still virgin country a people full of youth and life, receive it; it grows under the influence of the Christian principle of unity and brotherhood. A common faith unites all the members of that society of the middle ages, so strangely broken up; those nations, so different, so hostile to each other in appearance, nevertheless look upon one another as brothers, and form together the great family of Christianity. The circle of civilization soon widens, and embraces all Europe in the same range of improvement; no-people, however, takes part unless it shares the common faith; but, from the day of its conversion, also dates its entrance upon the path of progress.

Meantime, through many internal struggles, great states are gradually forming, the modern nations appear; full a thousand years have scarcely sufficed for these protracted throes. Different in characters, opposite in interests, long isolated from each other, these nations, having grown to maturity, enter into reciprocal relations. These relations are hostile at first; but the blending of so many various interests hastens their progress; bonds of intimacy are established; a greater community of interests, of ideas, of civilization in a word, strengthens the craving for harmony, and the balance of power in Europe becomes the aim of all high policy.

This equilibrium of material forces is finally changed in the nineteenth century into a European concert. Europe gives to the world, for the first time, the spectacle of a family of states, so closely bound together that they are only different members of the same body. No longer united by material ties alone, they are already bound by spiritual ties. From the depths of Russia to the ends of England, from Sicily to Cape North, we find the same religion at the basis of the social condition of all nations. The old ideas are a common property; new ideas speed almost through this whole space with the rapidity of thought, and reach, at the same time, the understandings of all. The manners, customs, sentiments, become every day more alike; in all things, community and intimacy are closer and closer. Nothing that touches the smallest, the most remote of the members of the great confederacy remains foreign or indifferent to the whole.

And yet the assimilation of the people of Europe stops far short of confounding their distinctive qualities. Not long since, the world saw them, with some surprise perhaps, protesting against the complete fusion seemingly about to annihilate their individual existence, and threatening to carry them back to the chaos of a homogeneous unity. They have once again proclaimed the power of historical ties, uniting the offspring of the same people, like friends of childhood, by a long community of life, and the vitality of those elements of race which bear witness to the original diversity of the gifts the Creator has bestowed upon his children. Each of the great physical districts composing that continent, in reality sustains a people whose moral and intellectual character, aptitudes, talents, differ as much as their language from those of their brethren. Each of these nations plays, in the great drama of history, a special part in accordance with its particular gifts, and all together form, in truth and reality, one of those rich organic unities which we have recognized as being the natural result of all regular and healthy growth.

This variety of elements and their reciprocal influence, joined to the community of action, which is the distinctive feature of modern society, exalt the powers of man to a degree hitherto unknown. Christian Europe beholds poetry, the arts, and the sublimest sciences, successively flourish, as in the bright days of pagan Greece; but, enriched already with the spoils of the past, culture is far more comprehensive, more varied, more profound; for it is not only affluent with the wealth of days gone by, but Christianity has placed it on the solid foundation of truth. The spirit of investigation ranges in all directions; it adds to this brilliant crown a new gem, the science of nature, growing with a speed of which the Ancient World had not even a forecast. Unriddled by the spirit of man, nature has yielded up to him her secrets; her untiring forces are enlisted in the service of intellect, which knows how to guide their action for its own purposes. Who shall describe those thousand applications of the science of nature, those inventions of the arts, each more marvellous than the others, coming upon us with a daily surprise; those ingenious and mighty machines obeying without pause he orders of man, and under his watchful eye accomplishing, with the same ease, the most gigantic works and the most delicate operations? The ocean has lost its terrors; with the help of steam the sailor braves opposing winds and waves; the compass and the stars conduct him with unerring precision to the end of his voyage. Space is annihilated by railroads; the word of man, borne on the wings of electricity, outruns in its course the sun himself; distances vanish, obstacles are smoothed away. Man thus disposes at will with the forces of nature, and the earth at last serves her master.

Such is the spectacle presented by European civilization. Looking upon it only under this brilliant aspect, and in itself, the progress of man seems to be almost touching its final goal. Nevertheless, the plan traced by the Divine Founder of the Christian church is much more vast; the goal which He sets up is much higher. These precious gifts of culture are not to remain the exclusive property of a small number of privileged men, nor of a single society, of one continent alone; the Christian principle is broader; it is universal, like the love of Christ. An important work remains, then, to be done; the work of diffusion and of propagation. This work is two-fold; for it is a duty to extend to the greatest possible number of the members of the same community all the blessings of civilization, at the same time that it is a duty to help all the nations of the earth to enjoy them. The first is social; the second pertains to humanity in general. To bring them both about, European society must overpass its present boundaries.

Just as Greece, the model on a small scale of what all

Europe becomes on a large scale, imparted to the East and to Rome the civilization which was the fruit of her whole popular life, so Europe owes to the world both her sciences and her culture, and the gospel, her most priceless good. The realm of civilization, which has been gradually enlarging, must increase still further; it must have no other limits than those of the great globe itself.

All was prepared in nature and history to invite the society of Europe to take this glorious initiative, and to facilitate the task.

The position of the European continent, in the midst of the other continents, seems to destine it from the beginning to this important part; its situation on the shores of the ocean opens an easy access to the remotest countries.

The ocean is, in fact, gentlemen, the grand highway of the world; from the earliest ages the civilized nations, urged by a secret instinct of their coming destinies, seem tending unconsciously to gather themselves near its shores. Born on the banks of the great rivers of the East, they cluster afterwards round the Mediterranean, under the sway of Greece and Rome. The modern world exchanges this theatre, henceforth too narrow, for the basin of the oceans, and our ships sail over the vast expanse of waters with more ease and security than the triremes of Greece and Rome crossed their inland seas.

The progress made by man in Europe also renders him capable of undertaking this work. In that continent, so tractable in shape, so well made, so nicely adjusted to his forces, he has learned to subdue nature by intelligence, and has thrown off the yoke. The child of the East has become a man in the land of the West. Thus no obstacle dismays or arrests him; he sets forth, and, like the Rome of other days, the Europe of the present marches to the conquest of the world less by arms than by her colonists, her commerce, her civilization, and by the gospel, which she carries to all nations.

The first land her ships encounter is the New World, waiting, as we have seen, only for the active labors of the civilized races, to yield up to them all the treasures that lay unused in its bosom. The European nations bordering the Atlantic establish themselves there, and divide it among them. In North America, the people of the North of Europe — the Anglo-Saxons, the Germans, the French; in South America, the Spanish and the Portuguese. The contrast between the North and South, mitigated in the temperate regions of the mother country, is reproduced in the New World more strongly marked, and on a grander scale, between North America, with its temperate climate, its Protestant and progressive people, and South America, with its tropical climate, its Catholic and stationary inhabitants. The conquest of the New World was the fairest and the most useful the European communities could have made, both for themselves and for the accomplishment of their work. They are transported thither with all their means of action; they get the mastery of nature without exhausting efforts; they strike their roots deep in a receptive soil almost untenanted; and America, while preparing to make new advances in social science, is already laboring in concert with Europe for the civilization of the world, which will not be completed without her.

But Europe stops not here. The ocean still opens to her the way to the maritime countries, the most highly favored regions in every continent. Africa and Australia receive her colonists, who plant in that soil, rebellious to civilization, the habits and the manners of our communities; Asia herself, old and immovable Asia, the symbol of stability, is shaken to her very foundations. India beholds her political power crushed under the arms and by the skill of England, while Christianity and the light of knowledge undermine the ancient Brahmin edifice, threatening every day to bury beneath its ruins that Old World which has survived more than three thousand years. China, in her turn, is forced to open her gates, and the ideas of Christianity and civilization, together with the products of European industry, are piercing, little by little, into that old sanctuary of superstition. Finally, there is not, in the bosom of the oceans, an island so distant but that, with the visits of the ships from Christian lands, it receives some germs of future improvement. The work is everywhere preparing, or beginning to bear fruit; and instead of one of those invasions of barbarous hordes, which so often terrified the world, plunging it again, for whole centuries, into the darkness of ignorance, we gaze upon the magnificent and consoling spectacle of a peaceful but irresistible march of civilization, and of the light of knowledge to the conquest of the whole earth.

Certainly these are admirable beginnings, the harbingers of a still more brilliant future. But here is only a part of the work the Christian nations of modern Europe appear summoned to execute. To this spread of the blessings of civilization abroad, ought to correspond, as we have said, a work of diffusion within civilized society itself; to the *humanitary* work, a social work. The greatest possible number of the members,—all, if it may be, — each, according to the measure of his gifts and the position assigned him by Providence, ought to share in the well-being, in the light of knowledge, in the moral perfection, which are now the portion of but a few. These advantages should, at least, be placed within the reach of all those, who, by a wise activity, the first condition of all progress, render themselves worthy to receive such reward.

This progress, whereto at present all civilized society aspires, — this goal, towards which it is tending, instinctively urged on by the very principle that constitutes its life, — is shown, as from afar, by that beautiful formula, drawn from the gospel, but so shamefully perverted by the false friends of progress, that one hardly dares repeat it after them; it is proclaimed in words that are the motto of the present age — *Liberty*, *Equality*, *Fraternity*. Yes, gentlemen, liberty to unfold all the living forces, and all the good tendencies of man, but not his evil tendencies; the equality of rights lying in the moral nature of man, but not that absolute, impossible equality which is contrary to nature and the course of Providence, and annihilates all progress; that fraternity which is the law of the gospel, and substitutes, for the selfishness that isolates and kills, the fraternal love and devotion that unites and makes alive; a free people, a people of brethren; unobstructed individual growth, attended by all the advantages of social life; diversity in unity, — this is the dream of the existing world, this is the prayer expressed under every variety of form. I say the dream of the existing world; for the perfect realization of such an ideal is possible only in perfect obedience to the divine law, in absolute goodness. In this earthly state, man, the sinner, must content himself with tending towards it, and drawing nearer every day.

Europe, gentlemen, has conceived the idea, and commenced the execution of the work. If we cast a glance back upon the phases of her progress, we see that each step she has taken in culture is at the same time marked by an amelioration in the state of the lower classes of society. From epoch to epoch, instruction and well-being become more and more universal. But historical ties of every kind, ancient customs, acquired rights, as much to be respected as any other rights of man, and, above all, the want of resources and of room for an ever-increasing population, are almost insurmountable difficulties, seeming to indicate that the work begun upon her soil is to be finished elsewhere. In Europe, the present must take the past into account, and in her past, Europe has roots too deeply fixed to adapt herself readily to all the exigencies of a new principle. Cut off the roots of this tree, ancient, but still majestic, still flowing with sap, and you take away its life. Cut from it a shoot, set that shoot in a fresh and virgin soil, and a new tree, at once strong and flexible, will readily take whatever form the skilful gardener shall desire to give it. Is not this what has been done for modern society by Him who dresses the great garden of humanity?

Yes, gentlemen, a new work is preparing, and a grave question is propounded. To what people shall it belong to carry out this work into reality? The law of history replies, to a new people. And to what continent? The geographical march of civilization tells us, to a new continent west of the Old World — to America.

This conclusion may seem a bold one; for the future is still covered by a veil it were unwise to wish to lift. Nevertheless, many signs seem to authorize this anticipation. It is worth the trouble of marking their existence, and of seeking to understand their language.

What is that new people, forming and growing upon the land of the future?

Is it a new race? No; for the ties of race imposed by physical nature must disappear in that world of emancipation and of liberty, to leave all its spontaneous character to the activity of man.

Is it some particular nation of the Old World? No; for if one people seems to stamp the physiognomy, yet the historical nations of every language and of every character are flowing thither, and blending together in one and the same nationality. The historical walls of separation in the Old World have fallen at once, and without a struggle. The European, who sets foot on American ground, with the purpose of making it his country, throws aside, at the threshold, not his affections and his memories, but his social and political past — if I may say it, takes a fresh start, recommences a new existence. He is received, by those who have gone thither before him, as a brother, entitled to the same immunities they are themselves enjoying. The most varied elements are gathering and harmonizing in this American people, which is moulding itself as no other ever did before, and which, more than any other people, is preëminently the cosmopolite, by virtue of its very constitution.

And what is the vital principle we find at the very root of this nation? It is the gospel. Not the gospel disfigured and cramped by the iron fetters of a powerful hierarchical church, like that the Christian Germanic world received while in its cradle, but the gospel restored by the Reformation, with its life-giving doctrines, and its regenerative power. Luther drew the Bible forth from the dust of libraries, where it lay forgotten, at the moment when Columbus discovered the New World. Will any one believe that here was only an accidental coincidence? More than this, gentlemen; for the first foundations were then laid of the edifice rising at the present day before our eyes, the actual construction of which, three centuries and a half later, enables us to see the providential connection of the two events.

The founders of social order in America are indeed the true offspring of the Reformation, — true Protestants. The Bible is their code. Imbued with the principles of civil and religious liberty they find written in the gospel, and for which they have given up their former country, they put them in practice in this land of their choice. They are all brethren, children of the same

Father — this is equality, independence, liberty. They submit from the heart to their Divine Leader, and to his law; this is the principle of order. Now the union of these two terms is free obedience to the divine will, which is the condition of a normal development, the supreme end of the education of man.

These, you will agree, gentlemen, are the sublime doctrines whence flow the religious, political, and social forms that distinguish America at the present time, from all the other countries on the globe. In religion, as in politics, democracy; the principle of free association pervading every part of public and private life; the preponderance of the judicial element set above the state itself, as the divine law is placed above human liberty, free obedience to the law, finally, rendering the means of constraint almost superfluous, and guaranteeing at once both security and liberty; — these are so many Christian ideas that have been incorporated in society. so many blessings America will continue to enjoy in proportion as she shall be faithful to the great principles whence they emanate.

A last characteristic, finally, of the nation forming on the soil of America, — upon which we fix our attention, because it furnishes in fact the representative of all modern progress, — is the greater emancipation from the dominion of nature. European society is transported to the New World, with all the power of modern arts and industry, which it applies without let or hindrance upon a large scale. Man, the master, now explores its vast territory. A perpetual movement, a fever of locomotion, rages from one end of the continent to the other. The America uses things without allowing himself to be taken ca tive by them. We behold everywhere the free will of man overmastering nature, which has lost the power of stamping him with a local character, of separating the nation into distinct peoples. Local country, which had so great sway in the Old World, no longer exists, so to speak, beyond the limits of the city, itself an association determined by man's free will, and not by the force of external nature. The great social country wins all interest, and all affection; it overmatches entirely geographical country.

Such are the principal lineaments that give to this people a character peculiarly their own. By these features we recognize the people of the future; for all the tendencies, struggling hard to find a vent in European society, are realized without effort here, because they are the very foundations whereon all the social relations rest. It is to this people, then, that the full and entire development belongs in the course of the epoch now beginning.

And what continent is better adapted than the American, to respond to the wants of humanity in this phase of its history?

The nations of Europe might easily be drawn out and arrayed within its vast confines. Its fertile soil secures prosperity to all, in exchange for their labor. Its forests, its treasures of coal laid up in quantities surpassing everything of the kind to be found in any part of the globe, prepare an inexhaustible support, and allow a future extension of industr to a degree and in proportions unknown elsewhere.

The simplicity and unity of plan we have observed in its configuration, its extensive plains, navigable rivers, the extreme facility of communications universally, with no serious obstacle lying in the way, from one end to the other of the fruitful part of the continent, all invite the inhabitants to frequent connection, to never-ceasing intercourse and exchange, checking the formation of local nationalities, and favoring the maintenance of a national unity, by the assimilation of all the parts.

Thus we may, perhaps, foresee that the American Union, already the most numerous association of men that has ever existed voluntarily united under the same law, will be able hereafter to become, even within the limits of its present confines, a true social world, transcending in grandeur and unity the most impressive spectacles of human greatness the history of past ages holds up to our view.

Finally, the oceanic position of the American continent secures its commercial prosperity, and creates, at the same time, the means of influence upon the world. It commands the Atlantic by its ports, while Oregon and California open the route of the Pacific Ocean and the East. America, also, is so placed as to take an active part in the great work of the civilization of the world, so admirably begun by Europe.

As Greece, then, gave the ancient world instruction and culture, so Europe instructs and refines the modern world, and all mankind; and as Rome wrought out the social work of antiquity, America seems called to do the same service for modern times, and to build up in the

New World the social state of which the Old World dreamed.

But while Rome accomplished her task by brute force, made a mere outside work, and brought about only an imperfect fusion of the nations, America is doing hers by persuasion. Drawing to her the free will of the sons of all the races, she binds them by one faith, and is thus preparing a true brotherhood of man. The one had only gross material arms; the other has spiritual arms. Between the two lies the whole distance that separates the heathen from the Christian world, and the progress made during two thousand years.

And further, what is there in common between this new social world of America and that world dreamed of by the morbid imagination of frantic Utopians, who, denying Christianity and its saving doctrines, renounce the vital principle of modern societies; who talk of progress, but fetter individual liberty, which is its sinew; of fraternity, as if man without God did not always relapse into selfishness; and show, finally, by their abortive attempts, both the corruption of the heart of man, and his inability to do the work of reconstructing society, which Divine Providence, in its wisdom, has reserved to itself?

The new society ought to receive entire the inheritance of those which have gone before; for nothing good or beautiful should perish. It ought to be rooted in that living faith which nourishes the nations and keeps up in them the freshness of life; its instruments should be the sciences and industry; its ornaments, literature and the fine arts; its end, the happiness of all, by training them up to moral perfection, and by spreading the gospel throughout the world, to the glory of the Redeemer.

You see, gentlemen, this picture transports us into the future. *There* stands the goal, and we are only now at the starting point. But this lofty goal may serve as a guiding star for the present, to preserve it from losing its way. In what measure and through what perils it shall be given to mankind, and to America in particular, to attain it, is known to God alone, and future ages will teach the issue to the world; but what we do know is, that it will be in proportion as man shall be faithful to the law of his moral nature, which is the divine law itself.

Asia, Europe, and North America, are the three grand stages of humanity in its march through the ages. Asia is the cradle where man passed his infancy, under the authority of law, and where he learned his dependence upon a sovereign master. Europe is the school where his youth was trained, where he waxed in strength and knowledge, grew to manhood, and learned at once his liberty and his moral responsibility. America is the theatre of his activity during the period of manhood; the land where he applies and practises all he has learned, brings into action all the forces he has acquired, and where he is still to learn that the entire development of his being and his own happiness are possible only by willing obedience to the laws of his Maker.

Thus lives and prospers, under the protection of the Divine Husbandman, the great tree of humanity, which is to overshadow the whole earth. It germinates and sends up its strong trunk in the ancient land of Asia. Grafted with a nobler stalk, it shoots out new branches, it blossoms in Europe. In America only, it seems destined to bear all its fruits. In these three we behold at once, as in a vast picture, the past, the present, and the future.

We see, then, that at each great phasis of the history of humanity the real work of the epoch is accomplished on a different theatre, and the centre or principal nucleus of civilized societies changes its place in the course of the ages. But in pointing out the remarkable fact of this successive displacement, let us not forget to state at the same time another movement, a progressive movement of extension, no less evident, and of almost equal importance. At first we behold the Orient shine alone; but soon the Occident ascends, and assumes the sceptre of intellectual light, and Greece now draws with her into a new progress the better portion of the East. Rome succeeds, and by her conquests removes the boundaries of the civilized world, whereof Italy is the soul, to the uttermost limits of the West. The North in succession is added, and all Europe becomes in turn the centre of a new world, which breaks the barriers seemingly imposed on it by nature, to enlarge and expand itself beyond the oceans. The establishment of European civilization in the New World, which has more than doubled the territorial extent of the cultivated nations, prepares an epoch of aggrandizement more rapid still. The two Americas, situated between the other four continents, seem destined to become, in their turn, a new centre of action, or a point of support for the establishment of easy and more rapid relations with all the nations of the world, and the irresistible logic of facts passing under our eyes, compels us to believe that during the epoch which is preparing, the boundaries of the domain of the civilized world can only be those of the globe itself.

Before closing let us cast back a glance upon the long way we have travelled over. The geographical march of history must have convinced us, if I am not mistaken,—

1. That the three continents of the North are organized for the development of man, and that we may rightfully name them preëminently the historical continents.

2. That each of these three continents, by virtue of its very structure, and of its physical qualities, has a special function in the education of mankind, and corresponds to one of the periods of his development.

3. That in proportion as this development advances, and civilization is perfected, and gains in intensity, the physical domain it occupies gains in extent and the number of cultivated nations increases.

4. That the entire physical creation corresponds to the moral creation, and is only to be explained by it.

Such, it seems, is the result of the study we have been making, of the relations between nature and history. It is not, perhaps, without some surprise, that we behold privileged continents and races, continents and races almost unalterably smitten with a character of inferiority. And yet, why be surprised at this? Is it not the attribute of God to dispense his gifts to whom he will, and as he will? Do we not know that in every organism there are needed divers members, clothed with functions more or less exalted, but alike necessary? We shall see that this great contrast of the historical continents and the continents of the inferior races seems established by Providence as a standing invitation addressed to man, bidding him unfold a new activity, and exercise the virtue of self-devotion, one of the highest to which his moral nature can be called. For the law of contrasts in the order of nature is the law of love in the moral order.

The three continents of the South, outcasts in appearance,—can they have been destined to an eternal isolation, doomed never to participate in that higher life of humanity, the sketch of which we have traced? And shall those gifts nature bestows on them with lavish hand, remain unused? No, gentlemen, such a doom cannot be in the plans of God. But the races inhabiting them are captives in the bonds of all powerful nature; they will never break down the fences that sunder them from us. It is for us, the favored races, to go to them. Tropical nature cannot be conquered and subdued, save by civilized men, armed with all the might of discipline, intelligence, and of skilful industry. It is, then, from the northern continents that those of the south await their deliverance; it is by the help of the civilized men of the temperate continents that it shall be vouchsafed to the man of the tropical lands to enter into the movement of universal progress and improvement, wherein mankind should share.

The privileged races have duties to perform, proportioned to the gifts they possess. To impart to other nations the advantages constituting their own glory, is the only way of legitimating the possession of them. We owe to the inferior races the blessings and the comforts of civilization; we owe them the intellectual development they are capable of; above all, we owe them the gospel, which is our glory, and will be their salvation; and if we neglect to help them partake in all these blessings, God will some time call us to a strict account.

In this way, alone, will the inferior races be able to come forth from the state of torpor and debasement wherein they are plunged, and live the active life of the higher races. Then shall commence, or rather shall rise to its just proportions, the elaboration of the material wealth of the tropical regions, for the benefit of the whole world. The nations of the lower races, associated like brothers with the civilized man of the ancient Christian societies, and directed by his intelligent activity, will be the chief instruments. The whole world, so turned to use by man, will fulfil its destiny.

The three northern continents, however, seem made to be the leaders; the three southern, the aids. The people of the temperate continents will always be the men of intelligence, of activity, the brain of humanity, if I may venture to say so; the people of the tropical continents will always be the hands, the workmen, the sons of toil.

History seems to be advancing towards the realization of these hopes, towards the solution of this great contrast. Each northern continent has its southern continent near by, which seems more especially commended to its guardianship and placed under its influence. Africa is already European at both extremities; North America leans on South America, which is indebted to the example of the North for its own emancipation and its own institutions. Asia is gradually receiving into her bosom the Christian nations of Europe, who are transforming her character, and beginning thence to settle the destinies of Australia. Lastly, the Christian missions are organizing upon a larger and larger scale in the two leading maritime countries of the globe, England and America, to whom the dominion of the sea seems granted for this end; and by engrafting upon all the nations the vital principle of civilized societies, without which no real community can exist between them, are preparing and hastening the true brotherhood, the spiritual brotherhood, of the whole human race.

It is in this great union, foretold alike by the order of nature and by the gospel, that every continent, as well as every people, will have its special functions, and that we shall find the solution and the definitive aim of all the physical and historical contrasts we have been studying. Everything in nature is arranged for the accomplishment by man of the admirable designs of Providence for the triumph of the good; and if man were faithful to his destination, the whole world would appear as a sublime concert of nature and the nations, blending their voices into a lofty harmony in praise of the Creator.

We are touching upon the close of our course; we

are far distant, indeed, from the point whence we started. Nevertheless, we have arrived hither, I believe, by a natural and regular path. Before we separate, gentlemen, allow me to add a few words upon the spirit and method that have animated and directed our studies.

All is life for him who is alive; all is death for him who is dead. All is spirit for him who is spirit; all is matter for him who is nothing but matter. It is with the whole life and the whole intellect that we should study the work of Him who is life and intellect itself.

This work of the Supreme Intelligence—can it be otherwise than intelligent? The work of Him who is all life and all love—must it not be living and full of love?

How should we not find in our earth itself the realization of an intelligent thought, of a thought of love to man, who is the end and aim of all creation, and the bright consummate flower of this admirable organization?

Yes, certainly, it is so. Faith so teaches, inspiring us with this sentiment, vague still, yet profound. Science so teaches by a patient and long-continued study, reserving this sublime view as the sweetest reward for our labor. Faith, enlightened and expounded by science,—the union of faith and science,—is living, harmonious knowledge, is perfected faith, for it has become VISION.

I have sought, gentlemen, to introduce you to the living knowledge of our globe, in the modest measure it is given me to do it. In spite of the imperfection of this knowledge, of which I feel that I have only touched upon the margin, if you have followed me you have had one more intellectual experience, and you admire with me the Author of so fair a creation.

If your heart has felt the benevolent purposes that have throughout presided over these arrangements, if it is convinced that everything in nature and history is ordained to guide us to happiness by lifting us up to Him, then it is grateful, then it loves in turn.

If the heart admires and loves, it adores; and that is the only worship worthy of rational man, the only service his Maker asks and accepts at his hands.

VALUABLE WORKS.

Diary and Correspondence OF THE LATE AMOS LAWRENCE. Edited by his son, WM. R. LAWRENCE, M. D. Octavo, cloth, $1.25; also, royal 12mo. ed., cl., $1.00.

Kitto's Popular Cyclopædia OF BIBLICAL LITERATURE. 500 illustrations. One vol., octavo. 812 pages. Cloth, $3.

Intended for ministers, theological students, parents, Sabbath-school teachers, and the great body of the religious public.

Analytical Concordance of the Holy SCRIPTURES; or, The Bible presented under Classified Heads or Topics. By JOHN EADIE, D. D. Octavo, 836 pp. $3.

Dr. Williams' Works.

Lectures on the Lord's Prayer — Religious Progress — Miscellanies.

☞ Dr. Williams is a profound scholar and a brilliant writer. — *New York Evangelist.*

Modern Atheism. Considered under its forms of Pantheism, Materialism, Secularism, Development and Natural Laws. By JAMES BUCHANAN, D. D., LL. D. 12mo. Cloth, $1.25.

The Hallig: or the Sheepfold in the Waters. A Tale of Humble Life on the Coast of Schleswig. From the German, by Mrs. GEORGE P. MARSH. 12mo. Cl., $1.

The Suffering Saviour. By Dr. KRUMMACHER. 12mo. Cloth, $1.25.

Heaven. By JAMES WM. KIMBALL. 12mo.

Christian's Daily Treasury. Religious Exercise for every Day in the Year. By Rev. E. TEMPLE. 12mo. Cloth, $1.

Wayland's Sermons. Delivered in the Chapel of Brown Univ. 12mo. Cl., $1.00.

Entertaining and Instructive Works FOR THE YOUNG. Elegantly illustrated. 16mo. Cloth, gilt backs.

The American Statesman. Life and Character of Daniel Webster. — *Young Americans Abroad;* or Vacation in Europe. — *The Island Home;* or the Young Cast-aways. — *Pleasant Pages for Young People.* — *The Guiding Star.* — *The Poor Boy and Merchant Prince.*

THE AIMWELL STORIES. Resembling and quite equal to the "Rollo Stories." — *Christian Register.* By WALTER AIMWELL. *Oscar;* or the Boy who had his own way. — *Clinton;* or Boy-Life in the Country. — *Ella;* or Turning over a New Leaf. — *Whistler;* or The Manly Boy. — *Marcus;* or the Boy Tamer.

WORKS BY Rev. HARVEY NEWCOMB. *How to be a Lady.* — *How to be a Man.* — *Anecdotes for Boys.* — *Anecdotes for Girls.*

BANVARD'S SERIES OF AMERICAN HISTORIES. *Plymouth and the Pilgrims.* — *Romance of American History.* — *Novelties of the New World, and Tragic Scenes in the History of Maryland and the old French War.*

God Revealed in Nature and in CHRIST. By Rev. JAMES B. WALKER, Author of "The Philosophy of the Plan of Salvation." 12mo. Cloth, $1.

Philosophy of the Plan of Salvation. New enlarged edition. 12mo. Cloth, 75 c.

Christian Life; SOCIAL AND INDIVIDUAL. By PETER BAYNE. 12mo. Cloth, $1.25.

All agree in pronouncing it one of the most admirable works of the age.

Yahveh Christ; or the Memorial Name. By ALEX. MACWHORTER. With an Introductory Letter, by NATH'L W. TAYLOR, D. D., in Yale Theol. Sem. 16mo. Cloth, 60 c.

The Signet Ring, AND ITS HEAVENLY MOTTO. From the German. 16mo. Cl., 31 c.

The Marriage Ring; or How to Make Home Happy. 18mo. Cloth, gilt, 75 c.

Mothers of the Wise and Good. By JABEZ BURNS, D. D. 16mo, Cloth, 75 c.

☞ A sketch of the mothers of many of the most eminent men of the world.

My Mother; or Recollections of Maternal Influence. 12mo. Cloth, 75c.

The Excellent Woman, with an Introduction, by Rev. W. B. SPRAGUE, D. D. Splendid Illustrations. 12mo. Cloth, $1.

The Progress of Baptist Principles IN THE LAST HUNDRED YEARS. By T. F. CURTIS, Prof. of Theology in the Lewisburg University. 12mo. Cloth, $1.25.

Dr. Harris' Works.

The Great Teacher. — The Great Commission. — The Pre-Adamite Earth. — Man Primeval. — Patriarchy. — Posthumous Works, 4 volumes.

The Better Land; OR THE BELIEVER'S JOURNEY AND FUTURE HOME. By Rev. A. C. THOMPSON. 12mo. Cloth, 85 c.

Kitto's History of Palestine, from the Patriarchal Age to the Present Time. With 200 Illustrations. 12mo. Cloth, $1.25.

An admirable work for the Family, the Sabbath and week-day School Library.

The Priest and the Huguenot; or, PERSECUTION IN THE AGE OF LOUIS XV. From the French of L. F. BUNGENER. Two vols., 12mo. Cloth, $2.25.

This is not only a work of thrilling interest, but is a masterly Protestant production.

The Psalmist. A Collection of Hymns for the Use of Baptist Churches. By BARON STOW and S. F. SMITH. With a SUPPLEMENT, containing an Additional Selection of Hymns, by RICHARD FULLER, D. D., and J. B. JETER, D. D. Published in various sizes, and styles of binding.

This is unquestionably the best collection of Hymns in the English language.

☞ In addition to works published by themselves, they keep an extensive assortment of works in all departments of trade, which they supply at publishers' prices. ☞ They particularly invite the attention of Booksellers, Travelling Agents, Teachers, School Committees, Librarians, Clergymen, and professional men generally (to whom a liberal discount is uniformly made), to their extensive stock. ☞ To persons wishing copies of Text-books, for examination, they will be forwarded, per mail or otherwise, on the reception of *one half* the price of the work desired. ☞ Orders from any part of the country attended to with faithfulness and dispatch.

www.ingramcontent.com/pod-product-compliance
Lightning Source LLC
LaVergne TN
LVHW020213110826
845151LV00003B/695